THE TEMPLE TIGER

AND MORE MAN-EATERS OF KUMAON

JIM CORBETT

THE TEMPLE TIGER

And More Man-Eaters of Kumaon

Illustrated by
RAYMOND SHEPPARD

DELHI
OXFORD UNIVERSITY PRESS
CALCUTTA   CHENNAI   MUMBAI

Oxford University Press, Great Clarendon Street, Oxford OX2 6DP

Oxford   New   York
Athens   Auckland   Bangkok   Calcutta
Cape Town   Chennai   Dar es Salaam   Delhi
Florence   Hong Kong   Istanbul   Karachi
Kuala Lumpur   Madrid   Melbourne   Mexico City
Mumbai   Nairobi   Paris   Singapore
Taipei   Tokyo   Toronto

and associates in
Berlin   Ibadan

First published 1954
Oxford India Paperbacks 1988
Twelfth impression 1997

ISBN 0 19 562257 X

Printed at Rekha Printers Pvt. Ltd., New Delhi 110020
and published by Manzar Khan, Oxford University Press
YMCA Library Building, Jai Singh Road, New Delhi 110001

Contents

Other illustrated editions
by the same author

MAN-EATERS OF KUMAON

**THE MAN-EATING LEOPARD
OF RUDRAPRAYAG**

The Temple Tiger

I

I T is not possible for those who have never lived in the upper reaches of the Himalayas to have any conception of the stranglehold that superstition has on the people who inhabit that sparsely populated region. The dividing line between the superstitions of simple uneducated people who live on high mountains, and the beliefs of sophisticated educated people who live at lesser heights, is so faint that it is difficult to determine where the one ends and the other begins. If therefore you are tempted to laugh at the credulity of the actors in the tale I am going to tell, I would ask you to pause for a moment and try to define the difference between superstition as exemplified in my tale, and your beliefs in the faith you have been brought up in.

Shortly after the Kaiser's war Robert Bellairs and I were on a shooting trip in the interior of Kumaon and we camped one September evening at the foot of Trisul, where we were informed that 800 goats were sacrificed each year to the demon of Trisul. With us we had fifteen of the keenest and the most cheerful hillmen I have ever been associated with on *shikar*. One of these men, Bala Singh, a Garhwali, had been with me for years and had accompanied me on many expeditions. It was his pride and pleasure when on *shikar* to select and carry the heaviest of my loads and, striding at the head of the other men, enliven the march with snatches of song. Round the camp-fire at night the men always sang part-songs before going to sleep, and during that first night, at the foot of Trisul, the singing lasted longer than usual and was accompanied by clapping of hands, shouting, and the beating of tin cans.

It had been our intention to camp at this spot and explore the country round for *baral* and *thar*, and we were very surprised as we sat down to breakfast next morning to see our men making preparations to strike camp. On asking for an explanation we were told that the site we had camped on was not suitable: that it was damp; that the drinking water was bad; that fuel was difficult to get; and, finally, that there was a better site two miles away.

I had six Garhwalis to carry my luggage and I noticed that it was being made up into five head-loads, and that Bala Singh was sitting apart near the camp-fire with a blanket over his head and shoulders. After breakfast I walked over to him, and noted as I did so that all the other men had stopped work and were watching me very intently. Bala Singh saw me coming and made no attempt to greet me, which was very unusual, and to all my questions he returned the one answer —that he was not ill. That day we did our two-mile march in silence, Bala Singh bringing up the rear and moving like a man who was walking in his sleep, or who was under the influence of drugs.

It was now quite apparent that whatever had happened to Bala Singh was affecting the other fourteen men, for they were performing their duties without their usual cheerfulness, and all of them had a strained and frightened look on their faces. While the 40-lb. tent Robert and I shared was being erected I took my Garhwali servant Mothi Singh—who had been with me for twenty-five years—aside and demanded to be told what was wrong with Bala Singh. After a lot of hedging and evasive answers I eventually got Mothi Singh's story, which, when it came, was short and direct. 'While we were sitting round the camp-fire last night and singing,' Mothi Singh said, 'the demon of Trisul entered Bala Singh's mouth and he swallowed him.' Mothi Singh went on to say that they had shouted and beaten tin cans to try to drive the demon out of Bala Singh, but that they had not succeeded in doing so, and that now nothing could be done about it.

Bala Singh was sitting apart, with the blanket still draped over his head. He was out of earshot of the other men, so, going over to him, I asked him to tell me what had happened the previous night. For a long minute Bala Singh looked up at me with eyes full of distress, and then in a hopeless tone of voice he said: 'Of what use is it, Sahib, for me to tell you what happened last night, for you will not believe me.' 'Have I ever', I asked, 'disbelieved you?' 'No,' he said, 'no, you have never disbelieved me, but this is a matter that you will not understand.' 'Whether I understand it or not,' I said, 'I want you to tell me exactly what happened.' After a long silence Bala Singh said: 'Very well, Sahib, I will tell you what happened. You know that in our hill-songs it is customary for one man to sing the verse, and for all the other men present to join in the chorus. Well, while I was singing a verse of one of our songs last night the demon of Trisul jumped into my mouth, and though I tried to eject him, he slipped down my throat into my stomach. The other men saw my struggle with the demon, for the fire was burning brightly, and they tried to drive him away by shouting and beating tin cans; but,' he added with a sob, 'the demon would not go.' 'Where is the demon now?' I asked. Placing his hand on the pit of his stomach Bala Singh answered with great conviction, 'He is here, Sahib, here; I can feel him moving about.'

Robert had spent the day prospecting the ground to the west of our camp and had shot a *thar*, of which he had seen several. After dinner we sat long into the night reviewing the situation. We had planned for, and looked forward to, this shoot for many months. It had taken Robert seven days' and me ten days' hard walking to reach our shooting ground, and on the night of our arrival Bala Singh had swallowed the demon of Trisul. What our personal opinions were on this subject did not matter, but what did matter was that every man in camp was convinced that Bala Singh had a demon in his stomach, and they were frightened of him and were shunning his company. To carry on a month's shoot under these conditions was not possible, and Robert very reluctantly agreed with me that the only thing to be done was for me to return to Naini Tal with Bala Singh, while he carried on with the shoot alone. So next morning I packed up my things, and after an early breakfast with Robert, set off on my ten days' walk back to Naini Tal.

Bala Singh, a perfect specimen of a man of about thirty years of age, had left Naini Tal full of the joy of life; now he returned silent, with a strained look in his eyes, and with the appearance of one who had lost all interest in life. My sisters, one of whom had been a medical missionary, did all they could for him. Friends from far and near came to visit him, but he just sat at the door of his house never speaking unless spoken to. The Civil Surgeon of Naini Tal, Colonel Cooke, a man of great experience and a close friend of the family, came to visit Bala Singh at my request. His verdict after a long and painstaking examination was, that Bala Singh was in perfect physical condition, and that he could ascribe no reason for the man's apparent depression.

A few days later I had a brain-wave. There was in Naini Tal at that time a very eminent Indian doctor and I thought if I could get him to examine Bala Singh and, after he had done so, tell him about the demon and persuade him to assure Bala Singh that there was no demon in his stomach he would be able to cure him of his trouble, for in addition to being a Hindu the doctor was himself a hillman. My brain-wave, however, did not work out as I had hoped and antici-pated, for as soon as he saw the sick man the doctor appeared to get suspicious and when in reply to some shrewd questions he learnt from Bala Singh that the demon of Trisul was in his stomach, he stepped away from him hurriedly and, turn-ing to me, said 'I am sorry you sent for me, for I can do nothing for this man.'

There were two men from Bala Singh's village in Naini Tal. Next day I sent for them. They knew what was wrong with Bala Singh for they had come to see him several times, and at my request they agreed to take him home. Provided with funds the three men started on their eight days' journey next morning. Three weeks later the two men returned and made their report to me.

Bala Singh had accomplished the journey without any trouble. On the night of his arrival home, and while his relatives and friends were gathered round him, he had sud-denly announced to the assembly that the demon wanted to be released to return to Trisul, and that the only way this could be accomplished was for him to die. 'So', my in-formants concluded, 'Bala Singh just lay down and died, and next morning we assisted at his cremation.'

Superstition, I am convinced, is a mental complaint similar to measles in that it attacks an individual or a community while leaving others immune. I therefore do not claim any credit for not contracting, while living on the upper reaches of the Himalayas, the virulent type of superstition that Bala Singh died of. But though I claim I am not superstitious I can give no explanation for the experience I met with at the bungalow while hunting the Champawat tiger, and the scream I heard coming from the deserted Thak village. Nor can I give any explanation for my repeated failures while engaged in one of the most interesting tiger hunts I have ever indulged in, and which I shall now relate.

2

No one who has visited Dabidhura is ever likely to forget the view that is to be obtained from the Rest House built near the summit of 'God's Mountain' by one who, quite evidently, was a lover of scenery. From the veranda of the little three-roomed house the hill falls steeply away to the valley of the Panar river. Beyond this valley the hills rise ridge upon ridge until they merge into the eternal snows which, until the advent of aircraft, formed an impenetrable barrier between India and her hungry northern neighbours.

A bridle-road running from Naini Tal, the administrative headquarters of Kumaon, to Loharghat, an outlying sub-division on her eastern border, passes through Dabidhura, and a branch of this road connects Dabidhura with Almora. I was hunting the Panar man-eating leopard—about which I shall tell you later—in the vicinity of this latter road when I was informed by a road overseer, on his way to Almora, that the leopard had killed a man at Dabidhura. So to Dabidhura I went.

The western approach to Dabidhura is up one of the steepest roads in Kumaon. The object the man who designed this road had in view was to get to the top by the shortest route possible, and this he accomplished by dispensing with hairpin bends and running his road straight up the face of the eight-thousand-foot mountain. After panting up this road on a hot afternoon in April I was sitting on the veranda of the Rest House drinking gallons of tea and feasting my eyes on the breath-taking view, when the priest of Dabidhura came to see me. When two years previously I had been hunting the Champawat man-eater, I had made friends with this frail old man, who officiated at the little temple nestling in the shadow of the great rock that had made Dabidhura a place of pilgrimage, and for whose presence in that unusual place I shall hazard no guess. When passing the temple a few minutes earlier I had made the customary offering which had been acknowledged by a nod by the old priest who was at his devotions. These devotions finished, the priest had crossed the road that runs between the temple and the Rest House and accepting a cigarette sat down on the floor of the veranda with his back against the wall for a comfortable chat. He was a friendly old man with plenty of time on his hands, and as I had done all the walking I wanted to that day, we sat long into the evening chatting and smoking.

From the priest I learnt that I had been misinformed by the road overseer about the man alleged to have been killed at Dabidhura the previous night by the man-eater. The alleged victim, a herdsman on his way from Almora to a village south of Dabidhura, had been the priest's guest the previous night. After the evening meal the herdsman had elected, against the priest's advice, to sleep on the chabutra (platform) of the temple. Round

about midnight, when the rock was casting a shadow over the temple, the man-eater crept up and, seizing the man's ankles, attempted to drag him off the platform. Awakening with a yell, the man grabbed a smouldering bit of wood from the nearby fire, and beat off the leopard. His yell brought the priest and several other men to his rescue and the combined force drove the animal away. The man's wounds were not serious, and after they had received rough-and-ready treatment at the hands of the *bania*, whose shop was near the temple, the herdsman continued his journey.

On the evidence of the priest I decided to remain at Dabidhura. The temple and the *bania's* shop were daily visited by men from the surrounding villages. These men would spread the news of my arrival and—knowing where I was to be found—I would immediately be informed of any kills of human beings, or of animals, that might take place in the area.

As the old priest got up to leave me that evening I asked him if it would be possible for me to get some shooting in the locality, for my men had been without meat for many days and there was none to be purchased at Dabidhura. 'Yes,' he answered, 'there is the temple tiger.' On my assuring him that I had no desire to shoot his tiger he rejoined with a laugh, 'I have no objection, Sahib, to your *trying* to shoot this tiger, but neither you nor anyone else will ever succeed in *killing* it.' And that is how I came to hear of the Dabidhura temple tiger, which provided me with one of the most interesting *shikar* experiences I have ever had.

3

The morning following my arrival at Dabidhura I went down the Loharghat road to see if I could find any trace of the man-eater, or learn anything about it in the villages adjoining the road, for the leopard was alleged to have gone in that direction after its attack on the man at the temple. On my return to the Rest House for a late lunch I found a man in conversation with my servant. This man informed me he had learnt from the priest that I wanted to do some shooting and he said he could show me a *jarao*—the hillman's name for *sambhar*—with horns as big as the branches of an oak tree. Hill *sambhar* do on occasions grow very fine horns—one had been shot in Kumaon some time previously with horns measuring forty-seven inches—and as a big animal would not only provide my men with meat but would also provide a meat ration for all at Dabidhura, I told the man I would accompany him after lunch.

Some months previously I had been to Calcutta on a short visit, and one morning walked into Manton's, the gunmaker's shop. On a glass showcase near the door was a rifle. I was looking at the weapon when the manager, who was an old friend of mine, came up. He informed me that the rifle, a .275 by Westley Richards, was a new model which the makers were anxious to introduce on the Indian market for hill shooting. The rifle was a beauty and the manager had little difficulty in persuading me to buy it on the understanding that if it did not suit me I would be at liberty to return it. So when I set out with my village friend that evening to shoot his *jarao* with horns as big as the branches of an oak tree, I was carrying my brand-new rifle.

To the south of Dabidhura the hill is less steep than it is to the north and we had proceeded in this direction through oak and scrub jungle for about two miles when we came to a grassy knoll with an extensive view of the valley below. Pointing to a small

patch of grass—surrounded by dense jungle—on the left-hand side of the valley, my guide informed me that the *jarao* came out to graze on this patch of grass morning and evening. He further informed me that there was a footpath on the right-hand side of the valley which he used when on his way to or from Dabidhura, and that it was from this path he was accustomed to seeing the *jarao*. The rifle I was carrying was sighted to five hundred yards and guaranteed to be dead accurate, and as the distance between the path and the *jarao's* feeding ground appeared to be only about three hundred yards I decided to go down the path and wait for a shot.

While we had been talking I had noticed some vultures circling to our left front. On drawing my companion's attention to them he informed me there was a small village in a fold of the hill in that direction and suggested that the vultures were possibly interested in some domestic animal that had died in the village. However, he said we would soon know what had attracted the birds, for our way lay through the village. The 'village' consisted of a single grass hut, a cattle shed, and an acre or so of terraced fields from which the crops had recently been cut. On one of these fields, separated from the hut and cattle shed by a ten-foot wide rain-water channel, vultures were tearing the last shreds of flesh from the skeleton of some large animal. A man walked out of the hut as we approached and, after greeting us, asked where I had come from and when I had arrived. On my telling him that I had come from Naini Tal to try to shoot the man-eating leopard and that I had arrived at Dabidhura the previous day he expressed great regret at not having known of my arrival. 'For you could then', he said, 'have shot the tiger that killed my cow.' He went on to tell me that he had tethered his fifteen head of cattle on the field, on which the vultures were pulling about the skeleton, the previous night, to fertilize it, and that during the night a tiger had come and killed one of the cows. He had no firearms and as there was no one within reach to whom he could appeal to shoot the tiger, he had gone to a village where a man lived who had the contract for collecting hides and skins in that area. This man had removed the hide of the cow two hours before my arrival, and the vultures had then carried out their function When I asked the man whether he had known that there was a tiger in the locality and, if so, why he had tethered his cattle out in the open at night, he surprised me by saying there had always been a tiger on the Dabidhura hill, but that up to the previous night it had never molested cattle.

As I moved away from the hut the man asked me where I was going and when I told him I was going to try to shoot the *jarao* on the far side of the valley, he begged me to leave the *jarao* alone for the present and to shoot the tiger. 'My holding is small and the land poor, as you can see,' he said, 'and if the tiger kills my cows, on which I depend for a living, my family and I will starve.'

While we had been talking, a woman had come up the hill with a *gharra* of water on her head, followed a little later by a girl carrying a bundle of green grass, and a boy carrying a bundle of dry sticks: four people living on an acre or so of poor land and a few pints of milk—for hill cattle give little milk—sold to the *bania* at Dabidhura. Little wonder, then, that the man was so anxious for me to shoot the tiger.

The vultures had destroyed the kill. This did not matter, however, for there was no heavy cover near the field where the tiger could have lain up and seen the vultures at their work, so he would be almost certain to return, for he had not been disturbed at his feed the previous night. My guide was also keen on my trying to shoot the tiger in preference to his *jarao*, so, telling the two men to sit down, I set off to try to find out in which direction the tiger had gone, for there were no trees on which I could sit near the field, and it was my intention to intercept the tiger on its way back. The hill was criss-crossed with cattle paths but the ground was too hard to show pug marks, and after circling round the village twice I eventually tried the rainwater channel. Here on the soft

B

damp ground I found the pug marks of a big male tiger. These pug marks showed that the tiger had gone up the channel after his feed, so it was reasonable to assume that he would return the same way. Growing out of the bank, on the same side of the channel as the hut and about thirty yards from it, was a gnarled and stunted oak tree smothered with a wild rose creeper. Laying down the rifle I stepped from the bank on to the tree, which was leaning out over the channel, and found there was a reasonably comfortable seat on the top of the creeper.

Rejoining the two men at the hut I told them I was going back to the Rest House for my heavy rifle, a double-barrelled .500 express using modified cordite. My guide very sportingly offered to save me this trouble, so after instructing him I sat down with the villager at the door of his hut and listened to the tales he had to tell of a poor but undaunted man's fight against nature and wild animals, to keep a grass roof above his head. When I asked him why he did not leave this isolated place and try to make a living elsewhere, he said, simply, 'This is my home.'

The sun was near setting when I saw two men coming down the hill towards the hut. Neither of them had a rifle, but Bala Singh—one of the best men who ever stepped out of Garhwal, and of whose tragic death some years later I have already told you—was carrying a lantern. On reaching me Bala Singh said he had not brought my heavy rifle because the cartridges for it were locked up in my suitcase and I had forgotten to send the key. Well, the tiger would have to be shot with my new rifle, and it could not have a better christening.

Before taking my seat on the tree I told the

owner of the hut that my success would depend on his keeping his two children, a girl of eight and a boy of six, quiet, and that his wife would have to defer cooking the evening meal until I had shot the tiger, or until I decided the tiger was not coming. My instructions to Bala Singh were to keep the inmates of the hut quiet, light the lantern when I whistled, and then await my further orders.

The vesper songs of the multitude of birds in the valley were hushed as the red glow from the setting sun died off the hills. Twilight deepened and a horned owl hooted on the hill above me. There would be a short period of semi-darkness before the moon rose. The time had now come, and the inmates of the hut were as silent as the dead. I was gripping the rifle and straining my eyes on the ground under me when the tiger, who had avoided passing under my tree, arrived at his kill and was angry at what he found. In a low muttering voice he cursed the vultures who, though they had departed two hours earlier, had left their musky smell on the ground they had fouled. For two, three, possibly four minutes he continued to mutter to himself, and then there was silence. The light was getting stronger. Another few minutes and the moon rose over the brow of the hill, flooding my world with light. The bones picked clean by the vultures were showing white in the moonlight, and nowhere was the tiger to be seen. Moistening my lips, which excitement had dried, I gave a low whistle. Bala Singh was on the alert and I heard him ask the owner of the hut for a light from the fire. Through the crevices of the grass hut I saw a glimmer of light, which grew stronger as the lantern was lit. The light moved across the hut and Bala Singh pulled open the door and stood on the threshold awaiting my further orders. With the exception of that one low whistle I had made no sound or movement from the time I had taken my seat on the tree. And now, when I looked down, there was the tiger standing below me, in brilliant moonlight, looking over his right shoulder at Bala Singh. The distance between the muzzle of my rifle and the tiger's head was about five feet, and the thought flashed through my mind that the cordite would probably singe his hair. The ivory foresight of my rifle was on the exact spot of the tiger's heart—where I knew my bullet would kill him instantaneously—when I gently pressed the trigger. The trigger gave under the pressure, and nothing happened.

Heavens! How incredibly careless I had been. I distinctly remembered having put a clip of five cartridges in the magazine when I took my seat on the tree but quite evidently when I pushed the bolt home it had failed to convey a cartridge from the magazine into the chamber, and this I had omitted to observe. Had the rifle been old and worn it might still have been possible to rectify my mistake. But the rifle was new and as I raised the lever to draw back the bolt there was a loud metallic click, and in one bound the tiger was up the bank and out of sight. Turning my head to see how Bala Singh had reacted, I saw him step back into the hut and close the door.

There was now no longer any need for silence and as Bala Singh came up at my call, to help me off the tree, I drew back the bolt of the rifle with the object of unloading the magazine and, as I did so, I noticed that the extractor at the end of the bolt held a cartridge. So the rifle had been loaded after all and the safety-catch off. Why then had the rifle not fired when I pulled the trigger? Too late, I knew the reason. One of the recommendations stressed by Manton's manager when showing me the rifle was that it had a double pull off. Never having handled a rifle with this so-called improvement, I did not know it was necessary, after the initial pull had taken up the slack, to pull the trigger a second time to release the striker. When I explained the reason for my failure to Bala Singh he blamed himself, 'for', said he, 'if I had brought your heavy rifle *and* the suitcase this would not have happened.' I was inclined to agree with him at the time, but as the days went by I was not so sure that even with the heavy rifle I would have been able to kill the tiger that evening.

4

Another long walk next morning, to try to get news of the man-eater, and when I returned to the Rest House I was greeted by a very agitated man who informed me that the tiger had just killed one of his cows. He had been grazing his cattle on the far side of the valley from where I sat the previous evening, when a tiger appeared and killed a red cow that had calved a few days previously. 'And now', he said, 'the heifer calf will die, for none of my other cows are in milk.'

Luck had been with the tiger the previous evening but his luck could not last indefinitely, and for the killing of this cow he would have to die, for cattle are scarce in the hills and the loss of a milch cow to a poor man was a serious matter. The man had no anxiety about the rest of his small herd, which had stampeded back to his village, so he was willing to wait while I had a meal. At 1 p.m. we set out, the man leading, I on his heels, and two of my men following with material for making a *machan*.

From an open patch of ground on the hillside my guide pointed out the lay of the land. His cattle had been grazing on a short stretch of grass a quarter of a mile below the ridge, when the tiger, coming up from the direction of the valley, had killed his cow. The rest of the herd had stampeded up the hill and over the ridge to his village, which was on the far side. Our shortest way was across the valley and up the other side, but I did not want to risk disturbing the tiger, so we skirted round the head of the valley to approach from above the spot where the cow had been killed. Between the ridge over which the cattle had stampeded and the spot where they had been grazing, was more or less open tree jungle. The tracks of the running animals had bitten deep into the soft loamy earth, and it was easy to follow these tracks back to where they had started. Here there was a big pool of blood with a drag-mark leading away from it. The drag led across the hill for 200 yards to a deep and well-wooded ravine with a trickle of water in it. Up this ravine the tiger had taken his kill.

The cow had been killed at about 10 a.m. on open ground, and the tiger's first anxiety would have been to remove it to some secluded spot where it would be hidden from prying eyes. So he had dragged it up the ravine and, after depositing it in a place he knew of, he had, as his pug-marks showed, gone down the ravine into the valley below. In an area in which human beings and cattle are moving about it is unwise to predict where a tiger will be lying up, for the slightest disturbance may make him change his position. So, though the pug-marks led down the ravine, the three men and I very cautiously followed the drag up the ravine.

Two hundred yards below the ridge along which we had come, rain-water had scooped out a big hole in the hillside. Here the ravine started. The hole, which at the upper end had a sheer drop into it of fifteen feet, had been made many years previously and was now partly overgrown with oak and ash saplings ten to twelve feet tall. Between these saplings and the fifteen-foot drop was a small open space on which the tiger had deposited his kill. I could sympathize with the owner of the cow when he told me with tears in his eyes that the fine animal that lay dead before us had been bred by him, and that it was a special favourite. No portion of the animal had been touched, the tiger having evidently brought it here to eat at his leisure.

A place had now to be found in which to sit. There were several big oak trees on either side of the ravine, but none overlooked the kill and all of them were unclimbable. Thirty yards below the kill and on the left-hand side of the ravine was a small stout holly tree. The branches were growing out at right angles to the trunk, and six feet above ground there was a strong enough branch for me to sit on and another on which to rest my feet. The three men protested strongly against my sitting so close to the ground. However, there was no other suitable place for me to sit, so the holly tree it would

have to be. Before sending the men away I instructed them
to go to the hut where I had been the previous evening, and to
wait there until I called to them, or until I joined them. The
distance across the valley was about half a mile and though
the men would not be able to see either me or the kill, I was
able to see the hut through the leaves of the holly tree.

The men left me at 4 p.m. and I settled down on the holly
branch for what I anticipated would be a long wait, for the
hill faced west and the tiger would probably not be on the
move much before sundown. To the left my field of vision—
through the holly leaves—extended down the ravine for fifty
yards. In front I had a clear view into the ravine, which was
about ten feet deep and twenty feet wide, and of the hill facing
me on which there were outcrops of rock but no trees. To the
right I had a clear view up to the ridge but I could not see the
kill, which was hidden by the thick growth of saplings.
Behind me was a dense thicket of *ringals* which extended
down to the level of my tree and further helped to mask the
kill. The tiger after depositing his kill in the hole, made by
rain-water, had gone down the ravine and it was reasonable to
assume that when he returned he would come by the same
route. So I concentrated all my attention on the ravine, in-
tending to shoot the tiger when he was at right angles to me.
That I could kill him at that short range I had no doubt what-
soever and to make quite sure of getting in a second shot, if it
was necessary, I cocked both hammers of my rifle.

There were *sambhar*, *kakar*, and *langur* in the jungle and
a great number of pheasants, magpies, babblers, thrushes,
and jays, all of which call on seeing a member of the cat
family, so I thought I would receive ample warning of the
tiger's coming. But here I was wrong, for without having
heard a single alarm call, I suddenly heard the tiger at his kill.
After going down the ravine, possibly for a drink, the tiger
had skirted round the thicket of *ringals* and approached his
kill without passing me. This did not worry me unduly for
tigers are restless at a kill in daylight, and I felt sure that

sooner or later the tiger would show up on the open ground in front of me. He had been eating for about fifteen minutes, tearing off great chunks of flesh, when I caught sight of a bear coming along the crest of the hill from left to right. He was a great big Himalayan black bear, and was strolling along as though it did not matter to him how long he took to get from here to there. Suddenly he stopped, turned facing downhill, and lay flat. After a minute or two he raised his head, snuffed the wind, and again lay flat. The wind, as always in daylight in the hills, was blowing uphill and the bear had got the scent of flesh and blood, mingled with the scent of tiger. I was a little to the right of the kill, so he had not got my scent. Presently he got to his feet and, with bent legs and body held close to the ground, started to stalk the tiger.

It was a revelation to me in animal stalking to see that bear coming down the hill. He had possibly two hundred yards to go and though he was not built for stalking, as tigers and leopards are, he covered the distance as smoothly as a snake and as silently as a shadow. The nearer he got the more cautious he became. I could see the lip of the fifteen-foot drop into the hole, and when the bear got to within a few feet of this spot he drew himself along with belly to ground. Waiting until the tiger was eating with great gusto the bear very slowly projected his head over the lip of the hole and looked down, and then as slowly drew his head back. Excitement with me had now reached the stage when the whole of my body was trembling, and my mouth and throat were dry.

On two occasions I have seen Himalayan bears walk off with tigers' kills. On both occasions the tigers were not present. And on two occasions I have seen bears walk up to feeding leopards and, after shooing them off, carry the kills away. But on this occasion the tiger—and a big male at that—was present on his kill and, further, he was not an animal to be shooed away like a leopard. At the back of my mind was the thought that surely this bear would not be so foolish as to try to dispossess the king of the jungle of his kill. But that was just what the bear appeared to intend doing, and his opportunity came when the tiger was cracking a bone. Whether the bear had been waiting for this moment I do not know; anyway, while the tiger was crunching the bone, the bear drew himself to the edge and, gathering his feet under him, launched himself into the hole with a mighty scream. The object of the scream I imagine was to intimidate the tiger, but so far from having this effect it appeared to infuriate him, for the bear's mighty scream was answered by an even mightier roar from the tiger.

Fights in the wild are very rare and this is only the second case I know of different species of animals fighting for the sake of fighting and not for the purpose of one using the other as food. I did not see the fight, for the reasons I have given, but I heard every detail of it. Waged in a hollow of restricted area the sound was terrifying and I was thankful that the fight was a straight one between two contestants who were capable of defending them-selves, and not a three-cornered one in which I was involved. Time stands still when every drop of blood racing through a rapidly beating heart is tingling with excitement. The fight may have lasted three minutes, or it may have lasted longer. Anyway, when the tiger considered he had administered

sufficient chastisement he broke off the engagement and came along the open ground in front of me at a fast gallop, closely followed by the still screaming bear. Just as I was aligning the sights of my rifle on the tiger's left shoulder he turned sharp to the left and, leaping the twenty-foot-wide ravine, landed at my feet. While he was still in the air I depressed the muzzle of the rifle and fired, as I thought, straight into his back. My shot was greeted with an angry grunt as the tiger crashed into the *ringals* behind me. For a few yards he carried on and then there was silence; shot through the heart and died in his tracks, I thought.

A .500 modified cordite rifle fired anywhere makes a con-

siderable noise, but here, in the ravine, it sounded like a
cannon. The detonation, however, had not the least effect on
the maddened bear. Following close on the heels of the tiger
he did not attempt to leap the ravine, as the tiger had done.
Storming down one bank he came up the other straight
towards me. I had no wish to shoot an animal that had the
courage to drive a tiger off his kill, but to have let that
screaming fury come any nearer would have been madness,
so, when he was a few feet from me, I put the bullet of the
left barrel into his broad forehead. Slowly he slid down the
bank on his stomach, until his haunches met the opposite
bank.

Where a moment earlier the jungle had resounded with
angry strife and the detonations of a heavy rifle, there was
now silence, and when my heart had resumed its normal beat,
my thoughts turned to a soothing smoke. Laying the rifle
across my knees I put both hands into my pockets to feel for
cigarette case and matches. At that moment I caught sight
of a movement on my right and, turning my head, saw the
tiger unhurriedly cantering along on the open ground over
which he had galloped a minute or two earlier and looking
not at me, but at his dead enemy.

I know that in relating these events as they occurred,
sportsmen will accuse me of rank bad shooting and gross
carelessness. I have no defence to make against the accusa-
tion of bad shooting, but I do not plead guilty to carelessness.
When I fired, as I thought, into the tiger's back, I was convinced I was delivering a fatal wound, and the angry response
followed by the mad rush and sudden cessation of sound was
ample justification for thinking the tiger had died in his tracks.
My second shot had killed the bear outright so there was no
necessity—while I was still on the tree—to reload the rifle
before laying it across my knees.

Surprise at seeing the tiger alive and unhurt lost me a
second or two, and thereafter I acted quickly. The rifle was
of the under-lever model; the lever being held in position by

two lugs on the trigger guard. This made the rapid loading of the rifle difficult, and, further, the spare cartridges were in my trousers pocket; easy to get at when standing up, but not so easy when sitting on a thin branch. Whether the tiger knew the bear was dead, or whether he was just keeping an eye on it to avoid a flank attack, I do not know. Anyway, he carried on across the face of the steep hill at a slow canter and had reached a spot forty yards away—which I can best describe as eleven o'clock—and was passing a great slab of rock when, with only one barrel loaded, I put up the rifle and fired. At my shot he reared up, fell over sideways, made a bad landing, scrambled to his feet, and cantered on round the shoulder of the hill with his tail in the air. The nickel-cased soft-nosed bullet with steel base had struck the rock a few inches from the tiger's face and the blow-back had thrown him off his balance but had done him no harm.

After a quiet smoke I stepped down from the holly tree and went to have a look at the bear, who, I found, was even bigger than I had at first thought. His self-sought fight with the tiger had been a very real one, for blood from a number of deep cuts was seeping through the thick fur on his neck and in several places his scalp was torn right down to the bone. These wounds in themselves would have mattered little to a tough animal like a bear, but what did matter and what had annoyed him was the injury to his nose. All males resent being struck on the nose, and not only had the bear been struck on that tender spot but insult had been added to injury by his nose being torn in half. Reason enough for him to have chased the tiger with murder in his eyes, and for him to have ignored the report of my heavy rifle.

There was not sufficient time for me to call up my men to skin the bear, so I set off to collect them at the hut and get back to the Rest House before nightfall, for somewhere in that area there was a man-eater. My men, and the dozen or so villagers who had collected at the hut, were too intent on gazing across the valley to observe my approach, and when I walked in among them, they were dumb with amazement. Bala Singh was the first to recover speech, and when I heard his story I was not surprised that the assembled men had looked at me as one returned from the dead. 'We advised you', Bala Singh said, 'not to sit so close to the ground, and when we heard your first scream, followed by the tiger's roar, we were convinced that you had been pulled out of the tree and that you were fighting for your life with the tiger. Then, when the tiger stopped roaring and you continued to scream, we thought the tiger was carrying you away. Later we heard two reports from your rifle, followed by a third, and we were greatly mystified, for we could not understand how a man who was being carried away by a tiger could fire his rifle. And while we were consulting with these men what we should do, you suddenly appeared and we became speechless.' To men keyed-up and listening for sounds from where a tiger was being sat up for, the scream of a bear could easily be mistaken for the scream of a human being, for the two are very similar and at a distance would not be distinguishable from one another.

Bala Singh got a cup of tea ready for me while I told the men about the fight they had heard and about the bear I had shot. Bear's fat is greatly valued as a cure for rheumatism, and the men were delighted when I told them I did not want the fat and that they could share it with their friends. Next morning I set out to skin the bear, accompanied by a crowd of men who were anxious not only to get a share of the fat but also to see the animal that had fought a tiger. I have never measured or weighed a bear but have seen quite a few, and the one I skinned that morning was the biggest and the fattest Himalayan bear I have ever seen. When the fat and the other prized parts of the bear had been divided, a very happy throng of men turned their faces to Dabidhura, and the happiest and the most envied of all was Bala Singh who proudly carried, strapped to his back, the bear's skin I had given him.

The tiger did not return to finish his interrupted meal, and by evening the vultures had picked clean the bones of the cow and the bear.

5

Skinning a bear encased in fat is a very messy job, and as I plodded back to the Rest House for a hot bath and a late breakfast I met a very agitated forest guard, whose head-quarters were at Dabidhura. He had been at an outlying beat the previous night and on his return to Dabidhura that morning had heard at the *bania's* shop about the bear I had shot. Being in urgent need of bear's fat for his father, who was crippled with rheumatism, he was hurrying to try to get a share of the fat when he ran into a herd of stampeding cattle, followed by a boy who informed him that a tiger had killed one of his cows. The forest guard had a rough idea where the cattle had been grazing when attacked by the tiger, so while Bala Singh and the other men carried on to Dabidhura I set off with him to try to find the kill. Uphill and downhill we went for two miles or more until we came to a small valley. It was in this valley that the forest guard thought the cow had been killed.

There had been an auction of condemned stores at the Gurkha depot at Almora a few days previously, and my com-panion had treated himself to a pair of Army boots many sizes too big for his feet. In these he had clumped ahead of me until we came to the lip of the valley. Here I made him remove his boots and when I saw the condition of his feet I marvelled that a man who had gone barefoot all his life had, for the sake of vanity, endured such torture. 'I bought big boots', he told me, 'because I thought they would shrink.'

The boat-shaped valley, some five acres in extent, was like a beautiful park dotted over with giant oak trees. On the side from which I approached it the ground sloped gently down and was free of bushes, but on the far side the hill went up more steeply with a few scattered bushes on it. I stood on the lip of the valley for a few minutes scanning every foot of ground in my field of vision without seeing anything suspicious, and then went down the grassy slope followed by the forest guard, now walking silently on bare feet. As I approached the flat ground on the floor of the valley I saw that the dead leaves and dry twigs over a considerable area had been scratched together, and piled into a great heap. Though no part of the cow was visible I knew that under this pile of dead matter the tiger had hidden his kill, and very foolishly I did not inform my companion of this fact, for he told me later that he did not know that tigers were in the habit of hiding their kills. When a tiger hides his kill it is usually an indication that he does not intend lying up near it, but it is not safe to assume this always. So, though I had scanned the ground before entering the valley, I again stood perfectly still while I had another look.

A little beyond the piled-up leaves and twigs the hill went up at an angle of forty-five degrees, and forty yards up the hillside there was a small clump of bushes. As I was looking at these I saw the tiger, who was lying on a small bit of flat ground with his feet towards me, turn over and present his back to me. I could see part of his head, and a three-inch-wide strip of his body from shoulder to hindquarters. A head shot was out of the question, and nothing would be gained by inflicting a flesh wound. I had the whole afternoon and evening before me and as the tiger would be bound to stand up sooner or later, I decided to sit down and wait on events. As I came to this decision I caught sight of a movement on my left, and on turning my head saw a bear coming stealthily up the valley towards the kill, followed by two half-grown cubs. The bear had evidently heard the tiger killing the cow and after giving the tiger time to settle down—as I have done on many occasions—she was now coming to investigate, for unless they have a special reason bears do not move about at midday. Had I been standing on the lip of the valley, instead of a few feet from the kill, I believe I should have witnessed a very interesting sight, for on finding the kill, which with their keen scent they would have had no difficulty in doing, the bears would have started to uncover it. This would have awakened the tiger and I cannot imagine that he would have relinquished his kill without a fight, and the fight would have been worth seeing.

The forest guard, who all this time had been standing quietly behind me seeing nothing but the ground at his feet, now caught sight of the bears and exclaimed in a loud voice, 'Dekho Sahib, bhalu, bhalu.' The tiger was up and away in a flash, but he had some twenty yards of open ground to cover, and as I aligned my sights on him and pressed the trigger the forest guard, under the impression that I was facing in the wrong direction, grabbed my arm and gave it a jerk with the result that my bullet struck a tree a few yards from where I was standing. Losing one's temper anywhere does no good, least of all in a jungle. The forest guard, who did not know what the piled-up leaves implied and who had not seen the tiger, was under the impression that he had saved my life by drawing my attention to the dreaded bears, so there was nothing to be said. Alarmed by my shot the bears lumbered away while my companion urged me with a catch in his throat, 'Maro, maro!' ('Shoot, shoot!')

A very dejected forest guard walked back with me to Dabidhura, and to cheer him up I asked him if he knew of any place where I could shoot a *ghooral*, for my men were still without meat. Not only did the game little man know of such a place but he also volunteered, blisters and all, to take me to it. So after a cup of tea we set off accompanied by two of my men who, the forest guard said, would be needed to carry back the bag.

From the veranda of the Rest House the hill falls steeply away. Down this hill the forest guard led me for a few hundred yards until we came to a foot-wide *ghooral* track running across the face of the hill. I now took the lead and had proceeded for about half a mile to the right when, on

coming to a rocky ridge, I looked across a deep ravine and
saw a *ghooral* on the far side standing on a projecting rock
and looking into space, as all goats including *thar*, ibex, and
markor have a habit of doing. It was a male *ghooral*, as I
could see from the white disk on its throat, and the distance
between us was a shade over two hundred yards. Here now
was an opportunity not only of procuring meat for my men
but also of testing the accuracy of my new rifle. So, lying
down, I put up the two hundred-yard leaf sight and, taking
very careful aim, fired. At my shot the *ghooral* sank down on
the rock on which he had been standing, which was very
fortunate, for below him was a sheer drop of many hundred
feet. A second *ghooral*, which I had not seen, now ran up the
far side of the ravine followed by a small kid, and after
standing still and looking back at us several times carried on
round the shoulder of the hill.

While the forest guard and I had a smoke my two men set
off to retrieve the bag. Deprived of his share of the bear's fat
the forest guard was made happy by being promised a bit of
the *ghooral* and its skin, which he said he would make into a
seat for his father who, owing to age and rheumatism, spent
all his days basking in the sun.

6

A visit to the valley early next morning confirmed my suspicion that the tiger would not return to his kill, and that the
bears would. Little but the bones were left when the three
bears had finished with the kill, and that little was being
industriously sought for by a solitary king vulture when I
arrived on the scene.

The morning was still young so, climbing the hill in the
direction in which the tiger had gone the previous day, I
went over the ridge and down the far side to the Loharghat
road to look for tracks of the man-eating leopard. On my
return to the Rest House at midday I was informed of yet
another tiger kill. My informant was an intelligent young

c

man who was on his way to Almora to attend a court case and, being unable to spare the time to show me where he had seen the tiger killing a cow, drew a sketch for me on the floor of the veranda with a piece of charcoal. After a combined breakfast and lunch I set out to try to find the kill which—if the young man's sketch was correct—had taken place five miles from where I had fired at the tiger the previous day. The tiger, I found, had come on a small herd of cattle grazing on the banks of the stream flowing down the main valley and, judging from the condition of the soft ground, had experienced some difficulty in pulling down the victim he had selected. Killing a big and vigorous animal weighing six or seven hundred pounds is a strenuous job, and a tiger after accomplishing this feat usually takes a breather. On this occasion, however, the tiger had picked up the cow as soon as he had killed it—as the absence of blood indicated—and crossing the stream entered the dense jungle at the foot of the hill.

Yesterday the tiger had covered up his kill at the spot where he had done his killing, but today it appeared to be his intention to remove his kill to as distant a place as possible from the scene of killing. For two miles or more I followed the drag up the steep face of the densely wooded hill to where the tiger, when he had conveyed his heavy burden to within a few hundred yards of the crest, had got one of the cow's hind legs fixed between two oak saplings. With a mighty jerk *uphill*, the tiger tore the leg off a little below the hock, and leaving that fixed between the saplings went on with his kill. The crest of the hill at the point where the tiger arrived with his kill was flat and overgrown with oak saplings a foot or two feet in girth. Under these trees, where there were no bushes or cover of any kind, the tiger left his kill without making any attempt to cover it up.

I had followed the drag slowly, carrying only my rifle and a few cartridges; even so, when I arrived at the crest my shirt was wet and my throat dry. I could imagine, therefore, the thirst that the tiger must have acquired and his desire to quench it. Being in need of a drink myself I set out to find the nearest water, where there was also a possibility of finding the tiger. The ravine in which I had shot the bear was half a mile to the right and had water in it, but there was another ravine closer to the left and I decided to try that first.

I had gone down this ravine for the best part of a mile and had come to a place where it narrowed with steep shaly banks on either side when, on going round a big rock, I saw the tiger lying in front of me at a range of twenty yards. There was a small pool of water at this spot, and lying on a narrow strip of sand between the pool and the right-hand bank was the tiger. Here the ravine took a sharp turn to the right, and part of the tiger was on my side of the turn and part round the bend. He was lying on his left side with his back to the pool and I could see his tail and part of his hind legs. Between me and the sleeping animal was a great mass of dry branches that had been lopped from overhanging trees some time previously to feed buffaloes. It was not possible to negotiate this obstacle without making a noise, nor was it possible to go along either of the steep banks without causing small land-slides of shale. So the only thing to do was to sit down and wait for the tiger to give me a shot.

After his great exertion and a good drink, the tiger was sleeping soundly and for half an hour he made no movement. Then he turned on to his right side and a little more of his legs came into view. In this position he lay for a few minutes and then stood up, and withdrew round the bend. With finger on trigger I waited for him to reappear, for his kill was up the hill behind me. Minutes passed and then a *kakar* a hundred yards away went dashing down the hill barking hysterically, and a little later a *sambhar* belled. The tiger had gone; why, I did not know, for he had already taken as much exercise as any tiger needed to take, and it was not a case of his having scented me, for tigers have no sense of smell.

It did not matter, however, for presently he would return to the kill he had been at such pains to take to the top of the hill, and I would be there to receive him. The water in the pool where the tiger had drunk was ice-cold, and having slaked my thirst I was able to enjoy a long-deferred smoke.

The sun was near setting when I made myself comfortable on an oak tree ten yards to the east and a little to the right of the kill. The tiger would come up the hill from the west and it was not advisable to have the kill directly between us, for tigers have very good eyesight. From my seat on the tree I had a clear view of the valley and of the hills beyond; and when the setting sun, showing as a great ball of fire, was resting on the rim of the earth bathing the world in red, a *sambhar* belled in the valley below me. The tiger was on the move and there was plenty of time for him to arrive at his kill while there was still sufficient daylight for accurate shooting.

The ball of fire dipped below the horizon; the red glow died off the earth; twilight gave place to darkness; and all was silent in the jungle. The moon was in its third quarter, but the stars—nowhere more brilliant than in the Himalayas —were giving enough light for me to see the kill, which was white. The head of the kill was towards me and if the tiger came now and started to eat at the hindquarters I would not be able to see him, but by aligning my sights on the white kill and then raising the rifle and pressing the trigger, as the kill disappeared from view, there was a fifty-fifty chance of hitting the tiger. But here was no man-eater to be fired at under any conditions. Here was a 'temple' tiger who had never molested human beings and who, though he had killed four head of cattle in four consecutive days, had committed no crime against the jungle code. To kill him outright would benefit those who were suffering from his depredations, but to take an uncertain shot at night with the possibility of only wounding him and leaving him to suffer for hours, or if unrecovered to become a man-eater, was not justifiable in any circumstances.

Light was coming in the east, for the boles of the trees were beginning to cast vague shadows, and then the moon rose, flooding the open patches of the jungle with light. It was then that the tiger came. I could not see him but I knew he had come for I could feel and sense his presence. Was he crouching on the far side of the kill with just his eyes and the top of his head raised over the brow of the hill watching me? No, that was not possible, for from the time I had made myself comfortable on my seat I had become part of the tree and tigers do not go through a jungle scanning, without a reason, every tree they approach. And yet, the tiger was here, and he was looking at me.

There was sufficient light now for me to see clearly, and very carefully I scanned all the ground in front of me. Then as I turned my head to the right, to look behind, I saw the tiger. He was sitting on his hunkers in a patch of moonlight, facing the kill, with his head turned looking up at me. When he saw me looking down on him he flattened his ears, and as I made no further movement, his ears regained their upright position. I could imagine him saying to himself, 'Well, you have now seen me, and what are you going to do about it?' There was little I could do about it, for in order to get a shot I would have to turn a half-circle and it would not be possible to do this without alarming the tiger, who was looking at me from a range of fifteen feet. There was, however, just a possibility of my getting a shot from my left shoulder, and this I decided to try to do. The rifle was resting on my knees with the muzzle pointing to the left, and as I lifted it and started turning it to the right the tiger lowered his head and again flattened his ears. In this position he remained as long as I was motionless, but the moment I started to move the rifle again, he was up and away into the shadows behind.

Well, that was that, and the tiger had very definitely won

another round. As long as I sat on the tree he would not return, but if I went away he *might* come back and remove the kill; and as he could not eat a whole cow in a night I would have another chance next day.

The question that now faced me was where to spend the night. I had already walked some twenty miles that day and the prospect of doing another eight miles to the Rest House—through forest all the way—did not appeal to me. In any other locality I would have moved away from the kill for two or three hundred yards and slept peacefully on the ground, but in this locality there was a man-eating leopard, and man-eating leopards hunt at night. While sitting on the tree earlier in the evening, I had heard the distant pealing of cattle-bells, coming either from a village or from a cattle-station. I had pin-pointed the sound and I now set out to find where it had come from. Cattle-lifting is unknown in the Himalayas, and throughout Kumaon there are communal cattle-stations situated in the jungles close to the grazing grounds. I traced the bells I had heard to one of these stations, in which there were about a hundred head of cattle in a large open shed surrounded by a strong stockade. The fact that the station was in the depth of the jungle, and unguarded, was proof of the honesty of the hillfolk, and it was also proof that until my arrival cattle in the Dabidhura area had not been molested by tigers.

At night all animals in the jungle are suspicious, and if I was to spend the night under the protection of the inmates of the shed I would have to disarm their very natural suspicion. The tenants of our village at Kaladhungi keep about nine hundred head of cows and buffaloes, and having been associated with cattle from my earliest childhood I know the language they understand. Moving very slowly and speaking to the cattle I approached the shed, and on reaching the stockade sat down with my back to it to have a smoke. Several cows were standing near the spot where I sat and one of them now advanced and, putting its head through the bars of the stockade, started to lick the back of my head; a friendly gesture, but a wetting one, and here at an altitude of eight thousand feet the nights were cold. Having finished my cigarette, I unloaded my rifle and, covering it with straw, climbed the stockade.

Care was needed in selecting a place on which to sleep, for if there was an alarm during the night and the cattle started milling round it would be unsafe to be caught on the ground. Near the centre of the shed, and close to one of the roof-supports up which I could go if the need arose, there was an open space between two sleeping cows. Stepping over recumbent animals and moving the heads of standing ones to get past them, I lay down between the two that were lying back to back. There was no alarm during the night, so the necessity for me to shin up the roof-support did not arise, and with the warm bodies of the cows to keep off the night chills and with the honey-sweet smell of healthy cattle in my nostrils I slept as one at peace with all the world, tigers and man-eating leopards included.

The sun was just rising next morning when, on hearing voices, I opened my eyes and saw three men, armed with milking pails, staring at me through the bars of the stockade. The water I had drunk at the tiger's pool was all that had passed my lips since breakfast the previous day, and the warm drink of milk the men gave me—after they had recovered from their amazement at finding me asleep with their cattle—was very welcome. Declining the men's invitation to accompany them back to their village for a meal, I thanked them for their board and lodging and, before returning to the Rest House for a bath and a square meal, set off to see where the tiger had taken his kill. To my surprise I found the kill lying just as I had left it, and after covering it over with branches to protect it from vultures and golden-headed eagles I went on to the Rest House.

In no other part of the world, I imagine, are servants as tolerant of the vagaries of their masters as in India. When I returned to the Rest House after an absence of twenty-four hours, no surprise was expressed, and no questions asked. A hot bath was ready, clean clothes laid out, and within a very short time I was sitting down to a breakfast of porridge, scrambled eggs, hot *chapatis* and honey—the last a present from the old priest—and a dish of tea. Breakfast over, I sat on the grass in front of the Rest House admiring the gorgeous view and making plans. I had set out from my home in Naini Tal with one object, and one object only, to try to shoot the Panar man-eating leopard; and from the night it had tried to drag the herdsman off the temple platform nothing had been heard of it. The priest, the *bania*, and all the people in near and distant villages that I questioned informed me that there were occasions when for long periods the man-eater seemed to disappear off the face of the earth, and they were of the opinion that one of these periods had now started, but no one could say how long it would last. The area over which the man-eater was operating was vast, and in it there were possibly ten to twenty leopards. To find and shoot in that area one particular leopard—that had stopped killing human beings—without knowing where to look for it, was a hopeless job.

My mission as far as the man-eater was concerned had failed, and no useful purpose would be served by my prolonging my stay at Dabidhura. The question of the temple tiger remained. I did not feel that the killing of this tiger was any responsibility of mine; but I did feel, and felt very strongly, that my pursuit of him was inducing him to kill more cattle than he would otherwise have done. Why a male tiger started killing cattle on the day of my arrival at Dabidhura it was not possible to say, and whether he would stop when I went away remained to be seen. Anyway, I had tried my best to shoot him; had paid compensation for the damage he had done to the full extent of my purse; and he had provided me with one of the most interesting jungle experiences

I had ever had. So I harboured no resentment against him for having beaten me at every point in the exciting game we had played during the past four days. These four days had been very strenuous for me, so I would rest today and make an early start next morning on the first stage of my journey back to Naini Tal. I had just come to this decision when a voice from behind me said, 'Salaam, Sahib. I have come to tell you that the tiger has killed one of my cows.' One more chance of shooting the tiger, and whether I succeeded or not I would stick to my plan of leaving next morning.

7

Annoyed at the interference of human beings and bears the tiger had shifted his ground, and this last kill had taken place on the eastern face of the Dabidhura mountain several miles from where I had sat up for him the previous evening. The ground here was undulating, with patches of scrub and a few odd trees dotted here and there; ideal ground for *chukor* (hill partridge), but the last place in which I would have expected to find a tiger.

Running diagonally across the face of the mountain was a shallow depression. In this depression were patches of dense scrub, interspersed with open glades of short grass. At the edge of one of these glades the cow had been killed, dragged a few yards towards some bushes, and left in the open. On the opposite or downhill side of the glade to the kill, there was a big oak tree. On this tree, the only one for hundreds of yards round, I decided to sit.

While my men warmed a kettle of water for tea I scouted round to see if I could get a shot at the tiger on foot. The tiger I felt sure was lying up somewhere in the depression, but though I searched every foot of it for an hour I saw no sign of him.

The tree that was to provide me with a seat was leaning out towards the glade. Excessive lopping had induced a crop of small branches all up the trunk, which made the tree easy

to climb but obscured a view of the trunk from above.
Twenty feet up, a single big branch jutted out over the
glade, offering the only seat on the tree but not a comfortable one or one easy to reach. At 4 p.m. I sent my men away,
instructing them to go to a village farther up the hill and
wait for me, for I had no intention of sitting up after sundown.

The kill, as I have said, was lying in the open, ten yards
from me and with its hindquarters a yard or so from a dense
clump of bushes. I had been in position for an hour and was
watching a number of red-whiskered bulbuls feeding on a
raspberry bush to my right front, when on turning my eyes
to look at the kill I saw the tiger's head projecting beyond the
clump of bushes. He was evidently lying down, for his head
was close to the ground, and his eyes were fixed on me.
Presently a paw was advanced, then another, and then very
slowly and with belly to ground the tiger drew himself up to
his kill. Here he lay for several minutes without movement.
Then, feeling with his mouth, and with his eyes still fixed on
me, he bit off the cow's tail, laid it on one side and started to
eat. Having eaten nothing since his fight with the bear three
days before, he was hungry, and he ate just as a man would
eat an apple, ignoring the skin and taking great bites of flesh
from the hindquarters.

The rifle was across my knees pointing in the direction of
the tiger, and all I had to do was to raise it to my shoulder.
I would get an opportunity of doing this when he turned his
eyes away from me for a brief moment. But the tiger
appeared to know his danger, for without taking his eyes off
me he ate on steadily and unhurriedly. When he had consumed about fifteen or twenty pounds of flesh, and when the
bulbuls had left the raspberry bush and, joined by two black-
throated jays, were making a great chattering on the bushes
behind him, I thought it was time for me to act. If I raised
the rifle very slowly he would probably not notice the movement so, when the birds were chattering their loudest, I

started to do this. I had raised the muzzle possibly six inches, when the tiger slid backwards as if drawn back by a powerful spring. With rifle to shoulder and elbows on knees I now waited for the tiger to project his head a second time, and this I felt sure he would presently do. Minutes passed, and then I heard the tiger. He had skirted round the bushes and, approaching from behind, started to claw my tree where the thick growth of small branches on the trunk made it impossible for me to see him. Purring with pleasure the tiger once and again clawed the tree with vigour, while I sat on my branch and rocked with silent laughter.

I know that crows and monkeys have a sense of humour, but until that day I did not know that tigers also possessed this sense. Nor did I know that an animal could have the luck, and the impudence, that particular tiger had. In five days he had killed five cows, four of them in broad daylight. In those five days I had seen him eight times and on four occasions I had pressed a trigger on him. And now, after staring at me for half an hour and eating while doing so, he was clawing the tree on which I was sitting and purring to show his contempt of me.

When telling me of the tiger the old priest said: 'I have no objection, Sahib, to your *trying* to shoot this tiger but neither you nor anyone else will ever succeed in *killing* it.' The tiger was now, in his own way, confirming what the priest had said. Well, the tiger had made the last move in the exciting game we had played without injury to either of us, but I was not going to give him the satisfaction of having the last laugh. Laying down the rifle and cupping my hands I waited until he stopped clawing, and then sent a full-throated shout echoing over the hills which sent him careering down the hill at full speed and brought my men down from the village at a run. 'We saw the tiger running away with his tail in the air,' the men said when they arrived, 'and just see what he has done to the tree.'

Next morning I bade farewell to all my friends at Dabidhura, and assured them I would return when the man-eater got active again.

I visited Dabidhura several times in subsequent years, while hunting man-eaters, and I never heard of anyone having killed the temple tiger. So I hope that in the fulness of time this old warrior, like an old soldier, just faded away.

II

The Muktesar Man-Eater

EIGHTEEN miles to the north-north-east of Naini Tal is a hill eight thousand feet high and twelve to fifteen miles long, running east and west. The western end of the hill rises steeply and near this end is the Muktesar Veterinary Research Institute, where lymph and vaccines are produced to fight India's cattle diseases. The laboratory and staff quarters are situated on the northern face of the hill and command one of the best views to be had anywhere of the Himalayan snowy range. This range, and all the hills that lie between it and the plains of India, run east and west, and from a commanding point on any of the hills an uninterrupted view can be obtained not only of the snows to the north but also of the hills and valleys to the east and to the west as far as the eye can see. People who have lived at Muktesar claim that it is the most beautiful spot in Kumaon, and that its climate has no equal.

A tigress that thought as highly of the amenities of Muktesar as human beings did, took up her residence in the extensive forests adjoining the small settlement. Here she lived very happily on *sambhar*, *kakar* and wild pig, until she had the misfortune to have an encounter with a porcupine. In this encounter she lost an eye and got some fifty quills, varying in length from one to nine inches, embedded in the arm and under the pad of her right foreleg. Several of these quills after striking a bone had doubled back in the form of a

U, the point and the broken-off end being close together. Suppurating sores formed where she endeavoured to extract the quills with her teeth, and while she was lying up in a thick patch of grass, starving and licking her wounds, a woman selected this particular patch of grass to cut as fodder for her cattle. At first the tigress took no notice, but when the woman had cut the grass right up to where she was lying, the tigress struck once, the blow crushing in the woman's skull. Death was instantaneous, for, when found the following day, she was grasping her sickle with one hand and holding a tuft of grass, which she was about to cut when struck, with the other. Leaving the woman lying where she had fallen, the tigress limped off for a distance of over a mile and took refuge in a little hollow under a fallen tree. Two days later a man came to chip firewood off this fallen tree, and the tigress who was lying on the far side killed him also. The man fell across the tree, and as he had removed his coat and shirt and the tigress had clawed his back when killing him, it is possible that the sight of blood trickling down his body as he hung across the bole of the tree first gave her the idea that he was something that she could satisfy her hunger with. However that may be, before leaving him she ate a small portion from his back. A day later she killed her third victim deliberately, and without having received any provocation. Thereafter she became an established man-eater.

I heard of the tigress shortly after she started killing human beings, and as there were a number of sportsmen at Muktesar, all of whom were keen on bagging the tigress—who was operating right on their doorsteps—I did not consider it would be sporting of an outsider to meddle in the matter. When the toll of human beings killed by the tigress had risen to twenty-four, however, and the lives of all the people living in the settlement and neighbouring villages were endangered and work at the Institute slowed down, the veterinary officer in charge of the Institute requested Government to solicit my help.

My task, as I saw it, was not going to be an easy one, for, apart from the fact that my experience of man-eaters was very limited, the extensive ground over which the tigress was operating was not known to me and I therefore had no idea where to look for her.

Accompanied by a servant and two men carrying a roll of bedding and a suitcase, I left Naini Tal at midday and walked ten miles to the Ramgarh dak bungalow, where I spent the night. The dak bungalow *khansama* (cook, bottle-washer, and general factotum) was a friend of mine, and when he learnt that I was on my way to Muktesar to try to shoot the man-eater, he warned me to be very careful while negotiating the last two miles into Muktesar for, he said, several people had recently been killed on that stretch of the road.

Leaving my men to pack up and follow me I armed myself with a double-barrelled .500 express rifle using modified cordite, and making a very early start next morning arrived at the junction of the Naini Tal/Almora road with the Muktesar road just as it was getting light. From this point it was necessary to walk warily for I was now in the man-eater's country. Before zigzagging up the face of a very steep hill the road runs for some distance over flat ground on which grows the orange-coloured lily, the round hard seeds of which can be used as shot in a muzzle-loading gun. This was the first time I had ever climbed that hill and I was very interested to see the caves, hollowed out by wind, in the sandstone cliffs overhanging the road. In a gale I imagine these caves must produce some very weird sounds, for they are of different sizes and, while some are shallow, others appear to penetrate deep into the sandstone.

Where the road comes out on a saddle of the hill there is a small area of flat ground flanked on the far side by the Muktesar Post Office, and a small bazaar. The post office was not open at that early hour, but one of the shops was and the shopkeeper very kindly gave me directions how to find the dak bungalow, which he said was half a mile away on the northern face of the hill. There are two dak bungalows at Muktesar, one reserved for government officials and the other for the general public. I did not know this and my shopkeeper friend, mistaking me for a government official, possibly because of the size of my hat, directed me to the wrong one and the *khansama* in charge of the bungalow, and I, incurred the displeasure of the red tape brigade, the *khansama* by providing me with breakfast, and I by partaking of it. However, of this I was at the time happily ignorant, and later I made it my business to see that the *khansama* did not suffer in any way for my mistake.

While I was admiring the superb view of the snowy range, and waiting for breakfast, a party of twelve Europeans passed me carrying service rifles, followed a few minutes later by a sergeant and two men carrying targets and flags. The sergeant, a friendly soul, informed me that the party that had just passed was on its way to the rifle range, and that it was keeping together because of the man-eater. I learnt from the sergeant that the officer in charge of the Institute had received a telegram from Government the previous day informing him that I was on my way to Muktesar. The sergeant expressed the hope that I would succeed in shooting the man-eater for, he said, conditions in the settlement had become very difficult. No one, even in daylight, cared to move about alone, and after dusk everyone had to remain behind locked doors. Many attempts had been made to shoot the man-eater but it had never returned to any of the kills that had been sat over.

After a very good breakfast I instructed the *khansama* to tell my men when they arrived that I had gone out to try to get news of the man-eater, and that I did not know when I would return. Then, picking up my rifle, I went up to the post office to send a telegram to my mother to let her know I had arrived safely.

From the flat ground in front of the post office and the bazaar the southern face of the Muktesar hill falls steeply

D

away, and is cut up by ridges and ravines overgrown with dense brushwood, with a few trees scattered here and there. I was standing on the edge of the hill, looking down into the valley and the well-wooded Ramgarh hills beyond, when I was joined by the Postmaster and several shopkeepers. The Postmaster had dealt with the Government telegram of the previous day, and on seeing my signature on the form I had just handed in, he concluded I was the person referred to in the telegram and he and his friends had now come to offer me their help. I was very glad of the offer for they were in the best position to see and converse with everyone coming to Muktesar, and as the man-eater was sure to be the main topic of conversation where two or more were gathered together, they would be able to collect information that would be of great value to me. In rural India the post office and the *bania's* shop are to village folk what taverns and clubs are to people of other lands, and if information on any particular subject is sought, the post office and the *bania's* shop are the best places to seek it.

In a fold of the hill to our left front, and about two miles away and a thousand feet below us, was a patch of cultivation. This I was informed was Badri Sah's apple orchard. Badri, son of an old friend of mine, had visited me in Naini Tal some months previously and had offered to put me up in his guest house and to assist me in every way he could to shoot the man-eater. This offer, for the reason already given, I had not accepted. Now, however, as I had come to Muktesar at the request of the Government I decided I would call on Badri and accept his offer to help me, especially as I had just been informed by my companions that the last human kill had taken place in the valley below his orchard.

Thanking all the men who were standing round me, and telling them I would rely on them for further information, I set off down the Dhari road. The day was still young and before calling on Badri there was time to visit some of the villages farther along the hill to the east. There were no mile-

stones along the road, and after I had covered what I considered was about six miles and visited two villages I turned
back. I had retraced my steps for about three miles when I
overtook a small girl in difficulties with a bullock. The girl,
who was about eight years old, wanted the bullock to go in
the direction of Muktesar, while the bullock wanted to go in
the opposite direction, and when I arrived on the scene the
stage had been reached when neither would do what the
other wanted. The bullock was a quiet old beast, and with
the girl walking in front holding the rope that was tied round
his neck and I walking behind to keep him on the move he
gave no further trouble. After we had proceeded a short
distance I said:

'We are not stealing Kalwa, are we?' I had heard her
addressing the black bullock by that name.

'N—o,' she answered indignantly, turning her big brown
eyes full on me.

'To whom does he belong?' I next asked.

'To my father,' she said.

'And where are we taking him?'

'To my uncle.'

'And why does uncle want Kalwa?'

'To plough his field.'

'But Kalwa can't plough uncle's field by himself.'

'Of course not,' she said. I *was* being stupid, but then you
could not expect a Sahib to know anything about bullocks
and ploughing.

'Has uncle only got one bullock?' I next asked.

'Yes,' she said; 'he has only got one bullock now, but he did
have two.'

'Where is the other one?' I asked, thinking that it had
probably been sold to satisfy a debt.

'The tiger killed it yesterday,' I was told. Here was news
indeed, and while I was digesting it we walked on in silence,
the girl every now and then looking back at me until she
plucked up courage to ask:

'Have you come to shoot the tiger?'

'Yes,' I said, 'I have come to try to shoot the tiger.'

'Then why are you going away from the kill?'

'Because we are taking Kalwa to uncle.' My answer appeared to satisfy the girl, and we plodded on. I had got some very useful information, but I wanted more and presently I said:

'Don't you know that the tiger is a man-eater?'

'Oh, yes,' she said, 'it ate Kunthi's father and Bonshi Singh's mother, and lots of other people.'

'Then why did your father send you with Kalwa? Why did he not come himself?'

'Because he has *bhabari bokhar* [malaria].'

'Have you no brothers?'

'No. I had a brother but he died long ago.'

'A mother?'

'Yes, I have a mother; she is cooking the food.'

'A sister?'

'No, I have no sister.' So on this small girl had devolved the dangerous task of taking her father's bullock to her uncle, along a road on which men were afraid to walk except when in large parties, and on which in four hours I had not seen another human being.

We had now come to a path up which the girl went, the bullock following, and I bringing up the rear. Presently we came to a field on the far side of which was a small house. As we approached the house the girl called out and told her uncle that she had brought Kalwa.

'All right,' a man's voice answered from the house, 'tie him to the post, Putli, and go home. I am having my food.' So we tied Kalwa to the post and went back to the road. Without the connecting link of Kalwa between us, Putli (dolly) was now shy, and as she would not walk by my side I walked ahead, suiting my pace to hers. We walked in silence for some time and then I said:

'I want to shoot the tiger that killed uncle's bullock but I don't know where the kill is. Will you show me?'

'Oh, yes,' she said eagerly, 'I will show you.'

'Have you seen the kill?' I asked.

'No,' she said, 'I have not seen it, but I heard uncle telling my father where it was.'

'Is it close to the road?'

'I don't know.'

'Was the bullock alone when it was killed?'

'No, it was with the village cattle.'

'Was it killed in the morning or the evening?'

'It was killed in the morning when it was going out to graze with the cows.'

While talking to the girl I was keeping a sharp look-out all round, for the road was narrow and bordered on the left by heavy tree jungle, and on the right by dense scrub. We had proceeded for about a mile when we came to a well-used cattle track leading off into the jungle on the left. Here the girl stopped and said it was on this track that her uncle had told her father the bullock had been killed. I had now got all the particulars I needed to enable me to find the kill, and after seeing the girl safely to her home I returned to the cattle track. This track ran across a valley and I had gone along it for about a quarter of a mile when I came on a spot where cattle had stampeded. Leaving the track, I now went through the jungle, parallel to and about fifty yards below the track. I had only gone a short distance when I came on a drag-mark. This drag-mark went straight down into the valley and after I had followed it for a few hundred yards I found the bullock, from which only a small portion of the hindquarters had been eaten. It was lying at the foot of a bank about twenty feet high, and some forty feet from the head of a deep ravine. Between the ravine and the kill was a stunted tree, smothered over by a wild rose. This was the only tree within a reasonable distance of the kill on which I could sit with any hope of bagging the tiger, for there was no moon, and if the tiger came after dark—as I felt sure it would—the nearer I was to the kill the better would be my chance of killing the tiger.

It was now 2 p.m. and there was just time for me to call on Badri and ask him for a cup of tea, of which I was in need for I had done a lot of walking since leaving Ramgarh at four o'clock that morning. The road to Badri's orchard takes off close to where the cattle track joins the road, and runs down a steep hill for a mile through dense brushwood. Badri was near his guest house, attending to a damaged apple tree when I arrived, and on hearing the reason for my visit he took me up to the guest house which was on a little knoll overlooking the orchard. While we sat on the veranda waiting for the tea and something to eat that Badri had ordered his servant to prepare for me, I told him why I had come to Muktesar, and about the kill the young girl had enabled me to find. When I asked Badri why this kill had not been reported to the sportsmen at Muktesar, he said that owing to the repeated failures of the sportsmen to bag the tiger the village folk had lost confidence in them, and for this reason kills were no longer being reported to them. Badri attributed the failures to the elaborate preparations that had been made to sit over kills. These preparations consisted of clearing the ground near the kills of all obstructions in the way of bushes and small trees, the building of big *machans*, and the occupation of the *machans* by several men. Reasons enough for the reputation the tiger had earned of never returning to a kill. Badri was convinced that there was only one tiger in the Muktesar district and that it was slightly lame in its right foreleg, but he did not know what had caused the lameness, nor did he know whether the animal was male or female.

Sitting on the veranda with us was a big Airedale terrier. Presently the dog started growling, and looking in the direction in which the dog was facing, we saw a big *langur* sitting on the ground and holding down the branch of an apple tree, and eating the unripe fruit. Picking up a shotgun that was leaning against the railing of the veranda,

Badri loaded it with No. 4 shot and fired. The range was too great for the pellets, assuming any hit it, to do the *langur* any harm, but the shot had the effect of making it canter up the hill with the dog in hot pursuit. Frightened that the dog might come to grief, I asked Badri to call it back, but he said it would be all right for the dog was always chasing this particular animal, which he said had done a lot of damage to his young trees. The dog was now gaining on the *langur*, and when it got to within a few yards the *langur* whipped round, got the dog by the ears, and bit a lump off the side of its head. The wound was a very severe one, and by the time we had finished attending to it my tea and a plate of hot *puris* (unleavened bread fried in butter) was ready for me.

I had told Badri about the tree I intended sitting on, and when I returned to the kill he insisted on going with me accompanied by two men carrying materials for making a small *machan*. Badri and the two men had lived under the shadow of the man-eater for over a year and had no illusions about it, and when they saw that there were no trees near the kill—with the exception of the one I had selected—in which a *machan* could be built, they urged me not to sit up that night, on the assumption that the tiger would remove the kill and provide me with a more suitable place to sit up the following night. This was what I myself would have done if the tiger had not been a man-eater, but as it was I was disinclined to miss a chance which might not recur on the morrow, even if it entailed a little risk. There were bears in this forest and if one of them smelt the kill any hope I had of getting a shot at the tiger would vanish, for Himalayan bears are no respecters of tigers and do not hesitate to appropriate their kills. Climbing into the tree, smothered as it was by the rose bush, was a difficult feat, and after I had made myself as comfortable as the thorn permitted and my rifle had been handed up to me Badri and his men left, promising to return early next morning.

I was facing the hill, with the ravine behind me. I was

in clear view of any animal coming down from above, but
if the tiger came from below, as I expected, it would not
see me until it got to the kill. The bullock, which was white,
was lying on its right side with its
legs towards me, and at a distance
of about fifteen feet. I had taken
my seat at 4 p.m. and an hour later
a *kakar* started barking on the side
of the ravine two hundred yards
below me. The tiger was on the
move, and having seen it the *kakar*
was standing still and barking. For
a long time it barked and then it
started to move away, the bark growing fainter and fainter
until the sound died away round the shoulder of the hill. This
indicated that after coming within sight of the kill, the tiger
had lain down. I had expected this to happen after having
been told by Badri the reasons for the failures to shoot the
tiger over a kill. I knew the tiger would now be lying somewhere near by with his eyes and ears open, to make quite sure
there were no human beings near the kill, before he approached
it. Minute succeeded long minute; dusk came; objects on the
hill in front of me became indistinct and then faded out. I
could still see the kill as a white blur when a stick snapped
at the head of the ravine and stealthy steps came towards me,
and then stopped immediately below. For a minute or two
there was dead silence, and then the tiger lay down on the
dry leaves at the foot of the tree.

Heavy clouds had rolled up near sunset and there was now
a black canopy overhead blotting out the stars. When the
tiger eventually got up and went to the kill, the night could
best be described as pitch black. Strain my eyes as I would I
could see nothing of the white bullock, and still less of the
tiger. On reaching the kill the tiger started blowing on it.
In the Himalayas, and especially in the summer, kills attract
hornets, most of which leave as the light fades but those that

are too torpid to fly remain, and a tiger—possibly after bitter experience—blows off the hornets adhering to the exposed portion of the flesh before starting to feed. There was no need for me to hurry over my shot for, close though it was, the tiger would not see me unless I attracted its attention by some movement or sound. I can see reasonably well on a dark night by the light of the stars, but there were no stars visible that night nor was there a flicker of lightning in the heavy clouds. The tiger had not moved the kill before starting to eat, so I knew he was lying broadside on to me, on the right-hand side of the kill.

Owing to the attempts that had been made to shoot the tiger I had a suspicion that it would not come before dark, and it had been my intention to take what aim I could—by the light of the stars—and then move the muzzle of my rifle sufficiently for my bullet to go a foot or two to the right of the kill. But now that the clouds had rendered my eyes useless, I would have to depend on my ears (my hearing at that time was perfect). Raising the rifle and resting my elbows on my knees, I took careful aim at the sound the tiger was making, and while holding the rifle steady, turned my right ear to the sound, and then back again. My aim was a little too high, so, lowering the muzzle a fraction of an inch, I again turned my head and listened. After I had done this a few times and satisfied myself that I was pointing at the sound, I moved the muzzle a little to the right and pressed the trigger. In two bounds the tiger was up the twenty-foot bank. At the top there was a small bit of flat ground, beyond which the hill went up steeply. I heard the tiger on the dry leaves as far as the flat ground, and then there was silence. This silence could be interpreted to mean either that the tiger had died on reaching the flat ground or that it was unwounded. Keeping the rifle to my shoulder I listened intently for three or four minutes, and as there was no further sound I lowered the rifle. This movement was greeted by a deep growl from the top of the bank. So the tiger was unwounded, and had seen me. My seat on the tree had originally been about ten feet up but, as I had nothing solid to sit on, the rose bush had sagged under my weight and I was now possibly no more than eight feet above ground, with my dangling feet considerably lower. And a little above and some twenty feet from me a tiger that I had every reason to believe was a man-eater was growling deep down in his throat.

The near proximity of a tiger in daylight, even when it has not seen you, causes a disturbance in the blood stream. When the tiger is not an ordinary one, however, but a man-eater and the time is ten o'clock on a dark night, and you know the man-eater is watching you, the disturbance in the blood stream becomes a storm. I maintain that a tiger does not kill beyond its requirements, except under provocation. The tiger that was growling at me already had a kill that would last it for two or three days and there was no necessity for it to kill me. Even so, I had an uneasy feeling that on this occasion this particular tiger might prove an exception to the rule. Tigers will at times return to a kill after being fired at, but I knew this one would not do so. I also knew that in spite of my uneasy feeling I was perfectly safe so long as I did not lose my balance—I had nothing to hold on to—or go to sleep and fall off the tree. There was no longer any reason for me to deny myself a smoke, so I took out my cigarette case and as I lit a match I heard the tiger move away from the edge of the bank. Presently it came back and again growled. I had smoked three cigarettes, and the tiger was still with me, when it came on to rain. A few big drops at first and then a heavy downpour. I had put on light clothes when I started from Ramgarh that morning and in a few minutes I was wet to the skin, for there was not a leaf above me to diffuse the rain-drops. The tiger, I knew, would have hurried off to shelter under a tree or on the lee of a rock the moment the rain started. The rain came on at 11 p.m.; at 4 a.m. it stopped and the sky cleared. A wind now started to blow, to add to my discomfort, and where I had been cold before I was now

frozen. When I get a twinge of rheumatism I remember that night and others like it, and am thankful that it is no more than a twinge.

Badri, good friend that he was, arrived with a man carrying a kettle of hot tea just as the sun was rising. Relieving me of my rifle the two men caught me as I slid off the tree, for my legs were too cramped to function. Then as I lay on the ground and drank the tea they massaged my legs and restored circulation. When I was able to stand, Badri sent his man off to light a fire in the guest house. I had never previously used my ears to direct a bullet and was interested to find that I had missed the tiger's head by only a few inches. The elevation had been all right but I had not moved the muzzle of the rifle far enough to the right, with the result that my bullet had struck the bullock six inches from where the tiger was eating.

The tea and the half-mile walk up to the road took all the creases out of me, and when we started down the mile-long track to Badri's orchard wet clothes and an empty stomach were my only discomfort. The track ran over red clay which the rain had made very slippery. In this clay were three tracks: Badri's and his man's tracks going up, and the man's tracks going down. For fifty yards there were only these three tracks in the wet clay, and then, where there was a bend in the track, a tigress had jumped down from the bank on the right and gone down the track on the heels of Badri's man. The footprints of the man and the pug-marks of the tigress

showed that both had been travelling at a fast pace. There was nothing that Badri and I could do, for the man had a twenty-minute start of us, and if he had not reached the safety of the orchard he would long ere this have been beyond any help we could give him. With uneasy thoughts assailing us we made what speed we could on the slippery ground and were very relieved to find, on coming to a footpath from where the orchard and a number of men working in it was visible, that the tigress had gone down the path while the man had carried on to the orchard. Questioned later, the man said he did not know that he had been followed by the tigress.

While drying my clothes in front of a roaring wood-fire in the guest house, I questioned Badri about the jungle into which the tigress had gone. The path which the tigress had taken, Badri told me, ran down into a deep and densely wooded ravine which extended down the face of a very steep hill, for a mile or more, to where it was met by another ravine coming down from the right. At the junction of the two ravines there was a stream and here there was an open patch of ground which, Badri said, commanded the exit of both ravines. Badri was of the opinion that the tigress would lie up for the day in the ravine into which we had every reason to believe she had gone, and as this appeared to be an ideal place for a beat, we decided to try this method of getting a shot at the tigress, provided we could muster sufficient men to carry out the beat. Govind Singh, Badri's head gardener, was summoned and our plan explained to him. Given until midday, Govind Singh said he could muster a gang of thirty men to do the beat, and in addition carry out his master's orders to gather five *maunds* (four hundred and ten pounds) of peas. Badri had an extensive vegetable garden in addition to his apple orchard and the previous evening he had received a telegram informing him that the price of marrowfat peas on the Naini Tal market had jumped to four annas (four pence) a pound. Badri was anxious to take advantage of this good price and his men were gathering the peas to be dispatched by pack pony that night, to arrive in Naini Tal for the early morning market.

After cleaning my rifle and walking round the orchard, I joined Badri at his morning meal—which had been put forward an hour to suit me—and at midday Govind produced his gang of thirty men. It was essential for someone to supervise the pea-pickers, so Badri decided to remain and send Govind to carry out the beat. Govind and the thirty men were local residents and knew the danger to be apprehended from the man-eater. However, after I had told them what I wanted them to do, they expressed their willingness to carry out my instructions. Badri was to give me an hour's start to enable me to search the ravine for the tigress and, if I failed to get a shot, to take up my position on the open ground near the stream. Govind was to divide his men into two parties, take charge of one party himself, and put a reliable man in charge of the other. At the end of the hour Badri was to fire a shot and the two parties were to set off, one on either side of the ravine, rolling rocks down, and shouting and clapping their hands. It all sounded as simple as that, but I had my doubts, for I have seen many beats go wron

Going up the track down which I had come that morning, I followed the path that the tigress had taken, only to find after I had gone a short distance that it petered out in a vast expanse of dense brushwood. Forcing my way through for several hundred yards I found that the hillside was cut up by a series of deep ravines and ridges. Going down a ridge which I thought was the right-hand boundary of the ravine to be beaten, I came to a big drop at the bottom of which the ravine on my left met a ravine coming down from the right, and at the junction of the two ravines there was a stream. While I was looking down and wondering where the open ground was on which I was to take my stand, I heard flies buzzing near me and on following the sound found the remains of a cow that had been killed about a week before.

The marks on the animal's throat showed that it had been killed by a tiger. The tiger had eaten all of the cow, except a portion of the shoulders, and the neck and head. Without having any particular reason for doing so, I dragged the carcass to the edge and sent it crashing down the steep hill. After rolling for about a hundred yards the carcass fetched up in a little hollow a short distance from the stream. Working round to the left I found an open patch of ground on a ridge about three hundred yards from the hollow into which I had rolled the remains of the cow. The ground down here was very different from what I had pictured it to be. There was no place where I could stand to overlook the hillside that was to be beaten, and the tigress might break out anywhere without my seeing her. However, it was then too late to do anything, for Badri had fired the shot that was to let me know the beat had started. Presently, away in the distance, I heard men shouting. For a time I thought the beat was coming my way and then the sounds grew fainter and fainter until they eventually died away. An hour later I again heard the beaters. They were coming down the hill to my right, and when they were on a level with me I shouted to them to stop the beat and join me on the ridge. It was no one's fault that the beat had miscarried, for without knowing the ground and without previous preparation we had tried to beat with a handful of untrained men a vast area of dense brushwood that hundreds of trained men would have found it difficult to cope with.

The beaters had had a very strenuous time forcing their way through the brushwood, and while they sat in a bunch removing thorns from their hands and feet and smoking my cigarettes Govind and I stood facing each other, discussing his suggestion of carrying out a beat on the morrow in which every available man in Muktesar and the surrounding villages would take part. Suddenly, in the middle of a sentence, Govind stopped talking. I could see that something unusual had attracted his attention behind me, for his eyes

narrowed and a look of incredulity came over his face. Swinging round I looked in the direction in which he was facing, and there, quietly walking along a field that had gone out of cultivation, was the tigress. She was about four hundred yards away on the hill on the far side of the stream, and was coming towards us.

When a tiger is approaching you in the forest—even when you are far from human habitations—thoughts course through your mind of the many things that can go wrong to spoil your chance of getting the shot, or the photograph, you are hoping for. On one occasion I was sitting on a hillside overlooking a game track, waiting for a tiger. The track led to a very sacred jungle shrine known as *Baram ka Than*. Baram is a jungle God who protects human beings and does not permit the shooting of animals in the area he watches over. The forest in the heart of which this shrine is situated is well stocked with game and is a favourite hunting ground of poachers for miles round, and of sportsmen from all parts of India. Yet, in a lifetime's acquaintance with that forest, I do not know of a single instance of an animal having been shot in the vicinity of the shrine. When therefore I set out that day to shoot a tiger that had been taking toll of our village buffaloes, I selected a spot a mile from Baram's shrine. I was in position, behind a bush, at 4 p.m. and an hour later a *sambhar* belled in the direction from which I was expecting

the tiger. A little later and a little nearer to me a *kakar*
started barking; the tiger was coming along the track near
which I was sitting. The jungle was fairly open and consisted mostly of young *jamun* trees, two to three feet in girth.
I caught sight of the tiger—a big male—when he was two
hundred yards away. He was coming slowly and had reduced
the distance between us to a hundred yards when I heard the
swish of leaves, and on looking up saw that one of the *jamun*
trees whose branches were interlaced with another was beginning to lean over. Very slowly the tree heeled over until it
came in contact with another tree of the same species and of
about the same size. For a few moments the second tree
supported the weight of the first and then it, too, started to
heel over. When the two trees were at an angle of about
thirty degrees from the perpendicular they fetched up against
a third and smaller tree. For a moment or two there was a
pause, and then all three trees crashed to the ground. While
watching the trees, which were only a few yards from me,
I had kept an eye on the tiger. At the first sound of the leaves
he had come to a halt and when the trees crashed to the
ground he turned and, without showing any sign of alarm,
went back in the direction from which he had come. What
made the occurrence I had witnessed so unusual was that the
trees were young and vigorous; that no rain had fallen
recently to loosen their roots; that not a breath of air was
stirring in the forest; and, finally, that the trees had fallen
across the track leading to the shrine when the tiger had only
another seventy yards to cover to give me the shot I was
waiting for.

The chances of a shot being spoilt are greatly increased
when the quarry is in an inhabited area in which parties of
men may be travelling from one village to another or going
to or from markets, or where shots may be fired to scare away
langurs from apple orchards. The tigress still had three
hundred yards to go to reach the stream, and two hundred
yards of that was over open ground on which there was not a

E

single tree or bush. The tigress was coming towards us at a slight angle and would see any movement we made, so there was nothing I could do but watch her, and no tigress had ever moved more slowly. She was known to the people of Muktesar as the lame tiger, but I could see no sign of her being lame. The plan that was forming in my head as I watched her was to wait until she entered the scrub jungle, and then run forward and try to get a shot at her either before or after she crossed the stream. Had there been sufficient cover between me and the point the tigress was making for, I would have gone forward as soon as I saw her and tried either to get a shot at her on the open ground or, failing that, intercept her at the stream. But unfortunately there was not sufficient cover to mask my movements, so I had to wait until the tigress entered the bushes between the open ground and the stream. Telling the men not to move or make a sound until I returned, I set off at a run as the tigress disappeared from view. The hill was steep and as I ran along the contour I came to a wild rose bush which extended up and down the hill for many yards. Through the middle of the bush there was a low tunnel, and as I bent down to run through it my hat was knocked off, and raising my head too soon at the end of the tunnel I was nearly dragged off my feet by the thorns that entered my head. The thorns of these wild roses are curved and very strong and as I was not able to stop myself some embedded themselves and broke off in my head—where my sister Maggie had difficulty in removing them when I got home—while others tore through the flesh. With little trickles of blood running down my face I continued to run until I approached the hollow into which I had rolled the partly-eaten kill from the hill above. This hollow was about forty yards long and thirty yards wide. The upper end of it where the kill was lying, the hill above the kill, and the further bank, were overgrown with dense brushwood. The lower half of the hollow and the bank on my side were free of bushes. As I reached the edge of the

hollow and peered over, I heard a bone crack. The tigress
had reached the hollow before me and, on finding the old
kill, was trying to make up for the meal she had been
deprived of the previous night.

If after leaving the kill, on which there was very little flesh,
the tigress came out on to the open ground I would get a shot
at her, but if she went up the hill or up the far bank I would
not see her. From the dense brushwood in which I could
hear the tigress a narrow path ran up the bank on my side
and passed within a yard to my left, and a yard beyond the
path there was a sheer drop of fifty feet into the stream
below. I was considering the possibility of driving the tigress
out of the brushwood on to the open ground by throwing a
stone on to the hill above her, when I heard a sound behind
me. On looking round I saw Govind standing behind me
with my hat in his hand. At that time no European in India
went about without a hat, and having seen mine knocked off
by the rose-bush Govind had retrieved it and brought it to
me. Near us there was a hole in the hill. Putting my finger to
my lips I got Govind by the arm and pressed him into the
hole. Sitting on his hunkers with his chin resting on his
drawn-up knees, hugging my hat, he just fitted into the hole
and looked a very miserable object, for he could hear the
tigress crunching bones a few yards away. As I straightened
up and resumed my position on the edge of the bank, the
tigress stopped eating. She had either seen me or, what was
more probable, she had not found the old kill to her liking.
For a long minute there was no movement or sound, and then
I caught sight of her. She had climbed up the opposite bank,
and was now going along the top of it towards the hill. At
this point there was a number of six-inch-thick poplar
saplings, and I could only see the outline of the tigress as she
went through them. With the forlorn hope that my bullet
would miss the saplings and find the tigress I threw up my
rifle and took a hurried shot. At my shot the tigress whipped
round, came down the bank, across the hollow, and up the

path on my side, as hard as she could go. I did not know, at the time, that my bullet had struck a sapling near her head, and that she was blind of one eye. So what looked like a very determined charge might only have been a frightened animal running away from danger, for in that restricted space she would not have known from what direction the report of my rifle had come. Be that as it may, what I took to be a wounded and a very angry tigress was coming straight at me; so, waiting until she was two yards away, I leant forward and with great good luck managed to put the remaining bullet in the rifle into the hollow where her neck joined her shoulder. The impact of the heavy .500 bullet deflected her just sufficiently for her to miss my left shoulder, and her impetus carried her over the fifty-foot drop into the stream below, where she landed with a great splash. Taking a step forward I looked over the edge and saw the tigress lying submerged in a pool with her feet in the air, while the water in the pool reddened with her blood.

Govind was still sitting in the hole, and at a sign he joined me. On seeing the tigress he turned and shouted to the men on the ridge, 'The tiger is dead. The tiger is dead.' The thirty men on the ridge now started shouting, and Badri on hearing them got hold of his shot gun and fired off ten rounds. These shots were heard at Muktesar and in the surrounding villages, and presently men from all sides were converging on the stream. Willing hands drew the tigress from the pool, lashed her to a sapling and carried her in triumph to Badri's orchard. Here she was put down on a bed of straw for all to see, while I went to the guest house for a cup of tea. An hour later by the light of hand lanterns, and with a great crowd of men standing round, among whom were several sportsmen from Muktesar, I skinned the tigress. It was then that I found she was blind of one eye and that she had some fifty porcupine quills, varying in length from one to nine inches, embedded in the arm and under the pad of her right foreleg. By ten o'clock my job was finished, and

declining Badri's very kind invitation to spend the night with him I climbed the hill in company with the people who had come down from Muktesar, among whom were my two men carrying the skin. On the open ground in front of the post office the skin was spread out for the Postmaster and his friends to see. At midnight I lay down in the dak bungalow reserved for the public, for a few hours' sleep. Four hours later I was on the move again and at midday I was back in my home at Naini Tal after an absence of seventy-two hours.

The shooting of a man-eater gives one a feeling of satisfaction. Satisfaction at having done a job that badly needed doing. Satisfaction at having out-manoeuvred, on his own ground, a very worthy antagonist. And, greatest satisfaction of all, at having made a small portion of the earth safe for a brave little girl to walk on.

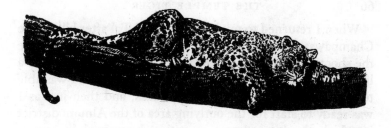

III

The Panar Man-Eater

I

WHILE I was hunting the Champawat man-eater in 1907 I heard of a man-eating leopard that was terrorizing the inhabitants of villages on the eastern border of the Almora district. This leopard, about which questions were asked in the House of Commons, was known under several names and was credited with having killed four hundred human beings. I knew the animal under the name of the Panar man-eater, and I shall therefore use this name for the purpose of my story.

No mention is made in Government records of man-eaters prior to the year 1905 and it would appear that until the advent of the Champawat tiger and the Panar leopard, man-eaters in Kumaon were unknown. When therefore these two animals—who between them killed eight hundred and thirty-six human beings—made their appearance Government was faced with a difficult situation for it had no machinery to put in action against them and had to rely on personal appeals to sportsmen. Unfortunately there were very few sportsmen in Kumaon at that time who had any inclination for this new form of sport which, rightly or wrongly, was considered as hazardous as Wilson's solo attempt—made a few years later—to conquer Everest. I myself was as ignorant of man-eaters as Wilson was of Everest and that I succeeded in my attempt, where he apparently failed in his, was due entirely to luck.

When I returned to my home in Naini Tal after killing the
Champawat tiger I was asked by Government to undertake
the shooting of the Panar leopard. I was working hard for a
living at the time and several weeks elapsed before I was able
to spare the time to undertake this task, and then just as I
was ready to start for the outlying area of the Almora district
in which the leopard was operating I received an urgent
request from Berthoud, the Deputy Commissioner of Naini
Tal, to go to the help of the people of Muktesar where a man-
eating tiger had established a reign of terror. After hunting
down the tiger, an account of which I have given, I went in
pursuit of the Panar leopard.

As I had not previously visited the vast area over which
this leopard was operating, I went via Almora to learn all I
could about the leopard from Stiffe, the Deputy Commissioner of Almora. He kindly invited me to lunch, provided me with maps, and then gave me a bit of a jolt when
wishing me goodbye by asking me if I had considered all the
risks and prepared for them by making my will.

My maps showed that there were two approaches to the
affected area, one via Panwanaula on the Pithoragarh road,
and the other via Lamgara on the Dabidhura road. I selected
the latter route and after lunch set out in good heart—despite
the reference to a will—accompanied by one servant and four
men carrying my luggage. My men and I had already done a
stiff march of fourteen miles from Khairna, but being young
and fit we were prepared to do another long march before
calling it a day.

As the full moon was rising we arrived at a small isolated
building which, from the scribbling on the walls and the
torn bits of paper lying about, we assumed was used as a
school. I had no tent with me and as the door of the building was locked I decided to spend the night in the courtyard
with my men, a perfectly safe proceeding for we were still
many miles from the man-eater's hunting grounds. This
courtyard, which was about twenty feet square, abutted on the

public road and was surrounded
on three sides by a two-foot-high
wall. On the fourth side it was
bounded by the school building.
There was plenty of fuel in the
jungle behind the school and my
men soon had a fire burning in a
corner of the courtyard for my
servant to cook my dinner on. I
was sitting with my back to the
locked door, smoking, and my
servant had just laid a leg of
mutton on the low wall nearest

the road and turned to attend to the fire, when I saw the head
of a leopard appear over the wall close to the leg of mutton.
Fascinated, I sat motionless and watched—for the leopard was
facing me—and when the man had moved away a few feet the
leopard grabbed the meat and bounded across the road into the
jungle beyond. The meat had been put down on a big sheet
of paper, which had stuck to it, and when my servant heard
the rustle of paper and saw what he thought was a dog running away with it he dashed forward shouting, but on
realizing that he was dealing with a leopard and not with a
mere dog he changed direction and dashed towards me with
even greater speed. All white people in the East are credited
with being a little mad—for other reasons than walking about
in the midday sun—and I am afraid my good servant thought
I was a little more mad than most of my kind when he found
I was laughing, for he said in a very aggrieved voice, 'It was
your dinner that the leopard carried away and I have nothing
else for you to eat.' However, he duly produced a meal that
did him credit, and to which I did as much justice as I am
sure the hungry leopard did to his leg of prime mutton.

Making an early start next morning, we halted at Lamgara
for a meal, and by evening reached the Dol dak bungalow

on the border of the man-eater's domain. Leaving my men at the bungalow I set out the following morning to try to get news of the man-eater. Going from village to village, and examining the connecting footpaths for leopard pug-marks, I arrived in the late evening at an isolated homestead consisting of a single stone-built slate-roofed house, situated in a few acres of cultivated land and surrounded by scrub jungle. On the footpath leading to this homestead I found the pug-marks of a big male leopard.

As I approached the house a man appeared on the narrow balcony and, climbing down a few wooden steps, came across the courtyard to meet me. He was a young man, possibly twenty-two years of age, and in great distress. It appeared that the previous night while he and his wife were sleeping on the floor of the single room that comprised the house, with the door open for it was April and very hot, the man-eater climbed on to the balcony and getting a grip of his wife's throat started to pull her head-foremost out of the room. With a strangled scream the woman flung an arm round her husband who, realizing in a flash what was happening, seized her arm with one hand and, placing the other against the lintel of the door, for leverage, jerked her away from the leopard and closed the door. For the rest of the night the man and his wife cowered in a corner of the room, while the leopard tried to tear down the door. In the hot unventilated room the woman's wounds started to turn septic and by morning her suffering and fear had rendered her unconscious.

Throughout the day the man remained with his wife, too frightened to leave her for fear the leopard should return and carry her away, and too frightened to face the mile of scrub jungle that lay between him and his nearest neighbour. As day was closing down and the unfortunate man was facing another night of terror he saw me coming towards the house, and when I had heard his story I was no longer surprised that he had run towards me and thrown himself sobbing at my feet.

A difficult situation faced me. I had not up to that time approached Government to provide people living in areas in which a man-eater was operating with first-aid sets, so there was no medical or any other kind of aid nearer than Almora, and Almora was twenty-five miles away. To get help for the woman I would have to go for it myself and that would mean condemning the man to lunacy, for he had already stood as much as any man could stand and another night in that room, with the prospect of the leopard returning and trying to gain entrance, would of a certainty have landed him in a mad-house.

The man's wife, a girl of about eighteen, was lying on her back when the leopard clamped its teeth into her throat, and when the man got a grip of her arm and started to pull her back the leopard—to get a better purchase—drove the claws of one paw into her breast. In the final struggle the claws ripped through the flesh, making four deep cuts. In the heat of the small room, which had only one door and no windows and in which a swarm of flies were buzzing, all the wounds in the girl's throat and on her breast had turned septic, and whether medical aid could be procured or not the chances of her surviving were very slight; so, instead of going for help, I decided to stay the night with the man. I very sincerely hope that no one who reads this story will ever be condemned to seeing and hearing the sufferings of a human being, or of an animal, that has had the misfortune of being caught by

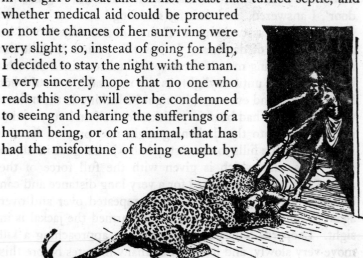

the throat by either a leopard or a tiger and not having the means—other than a bullet—of alleviating or of ending the suffering.

The balcony which ran the length of the house, and which was boarded up at both ends, was about fifteen feet long and four feet wide, accessible by steps hewn in a pine sapling. Opposite these steps was the one door of the house, and under the balcony was an open recess four feet wide and four feet high, used for storing firewood.

The man begged me to stay in the room with him and his wife but it was not possible for me to do this, for, though I am not squeamish, the smell in the room was overpowering and more than I could stand. So between us we moved the firewood from one end of the recess under the balcony, clearing a small space where I could sit with my back to the wall. Night was now closing in, so after a wash and a drink at a near-by spring I settled down in my corner and told the man to go up to his wife and keep the door of the room open. As he climbed the steps the man said, 'The leopard will surely kill you, Sahib, and then what will I do?' 'Close the door', I answered, 'and wait for morning.'

The moon was two nights off the full and there would be a short period of darkness. It was this period of darkness that was worrying me. If the leopard had remained scratching at the door until daylight, as the man said, it would not have gone far and even now it might be lurking in the bushes watching me. I had been in position for half an hour, straining my eyes into the darkening night and praying for the moon to top the hills to the east, when a jackal gave its alarm call. This call, which is given with the full force of the animal's lungs, can be heard for a very long distance and can be described as 'pheaon', 'pheaon', repeated over and over again as long as the danger that has alarmed the jackal is in sight. Leopards when hunting or when approaching a kill move very slowly, and it would be many minutes before this one—assuming it was the man-eater—covered the half mile between us, and even if in the meantime the moon had not risen it would be giving sufficient light to shoot by, so I was able to relax and breathe more freely.

Minutes dragged by. The jackal stopped calling. The moon rose over the hills, flooding the ground in front of me with brilliant light. No movement to be seen anywhere, and the only sound to be heard in all the world the agonized fight for breath of the unfortunate girl above me. Minutes gave way to hours. The moon climbed the heavens and then started to go down in the west, casting the shadow of the house on the ground I was watching. Another period of danger, for if the leopard had seen me he would, with a leopard's patience, be waiting for these lengthening shadows to mask his movements. Nothing happened, and one of the longest nights I have ever watched through came to an end when the light from the sun lit up the sky where, twelve hours earlier, the moon had risen.

The man, after his vigil of the previous night, had slept soundly and as I left my corner and eased my aching bones —only those who have sat motionless on hard ground for hours know how bones can ache—he came down the steps. Except for a few wild raspberries I had eaten nothing for twenty-four hours, and as no useful purpose would have been served by my remaining any longer, I bade the man goodbye and set off to rejoin my men at the Dol dak bungalow, eight miles away, and summon aid for the girl. I had only gone a few miles when I met my men. Alarmed at my long absence they had packed up my belongings, paid my dues at the dak bungalow, and then set out to look for me. While I was talking to them the Road Overseer, whom I have mentioned in my story of the Temple Tiger, came along. He was well mounted on a sturdy Bhootia pony, and as he was on his way to Almora he gladly undertook to carry a letter from me to Stiffe. Immediately on receipt of my letter Stiffe dispatched medical aid for the girl, but her sufferings were over when it arrived.

It was this Road Overseer who informed me about the human kill that took me to Dabidhura, where I met with one of the most interesting and the most exciting *shikar* experiences I have ever had. After that experience I asked the old priest of the Dabidhura temple if the man-eater had as effective protection from his temple as the tiger I had failed to shoot, and he answered, 'No, no, Sahib. This *shaitan* [devil] has killed many people who worshipped at my temple and when you come back to shoot him, as you say you will, I shall offer up prayers for your success morning and evening.'

2

No matter how full of happiness our life may have been, there are periods in it that we look back to with special pleasure. Such a period for me was the year 1910, for in that year I shot the Muktesar man-eating tiger and the Panar man-eating leopard, and in between these two—for me—great events, my men and I set up an all-time record at Mokameh Ghat by handling, without any mechanical means, five thousand five hundred tons of goods in a single working day.

My first attempt to shoot the Panar leopard was made in April 1910, and it was not until September of the same year that I was able to spare the time to make a second attempt. I have no idea how many human beings were killed by the leopard between April and September, for no bulletins were issued by Government and beyond a reference to questions asked in the House of Commons no mention of the leopard— as far as I am aware—was made in the Indian Press. The Panar leopard was credited with having killed four hundred human beings, against one hundred and twenty-five killed by the Rudraprayag leopard, and the fact that the former received such scant publicity while the latter was headline news throughout India was due entirely to the fact that the Panar leopard operated in a remote area far from the beaten track, whereas the Rudraprayag leopard operated in an area visited each year by sixty thousand pilgrims ranging from the humblest in the land to the highest, all of whom had to run the gauntlet of the man-eater. It was these pilgrims, and the daily bulletins issued by Government, that made the Rudraprayag leopard so famous, though it caused far less human suffering than the Panar leopard.

Accompanied by a servant and four men carrying my camp kit and provisions, I set out from Naini Tal on 10 September on my second attempt to shoot the Panar leopard. The sky was overcast when we left home at 4 a.m. and we had only gone a few miles when a deluge of rain came on. Throughout the day it rained and we arrived at Almora, after a twenty-eight-mile march, wet to the bone. I was to have spent the night with Stiffe, but not having a stitch of dry clothing to put on I excused myself and spent the night at the dak bungalow. There were no other travellers there and the man in charge very kindly put two rooms at my disposal, with a big wood fire in each, and by morning my kit was dry enough for me to continue my journey.

It had been my intention to follow the same route from Almora that I had followed in April, and start my hunt for the leopard from the house in which the girl had died of her wounds. While I was having breakfast a mason by the name of Panwa, who did odd jobs for us in Naini Tal, presented himself. Panwa's home was in the Panar valley, and on learning from my men that I was on my way to try to shoot the man-eater he asked for permission to join our party, for he wanted to visit his home and was frightened to undertake the journey alone. Panwa knew the country and on his advice I altered my plans and instead of taking the road to Dabidhura via the school where the leopard had eaten my dinner, I took the road leading to Pithoragarh. Spending the night at the Panwa Naula dak bungalow, we made an early start next morning and after proceeding a few miles left the Pithoragarh road for a track leading off to the right. We were now in the man-eater's territory where there were no roads, and where the only communication was along footpaths running from village to village.

Progress was slow, for the villages were widely scattered over many hundreds of square miles of country, and as the exact whereabouts of the man-eater were not known it was necessary to visit each village to make inquiries. Going through Salan and Rangot *pattis* (*patti* is a group of villages), I arrived late on the evening of the fourth day at Chakati, where I was informed by the headman that a human being had been killed a few days previously at a village called Sanouli on the far side of the Panar river. Owing to the recent heavy rain the Panar was in flood and the headman advised me to spend the night in his village, promising to give me a guide next morning to show me the only safe ford over the river, for the Panar was not bridged.

The headman and I had carried on our conversation at one end of a long row of double-storied buildings and when, on his advice, I elected to stay the night in the village, he said he would have two rooms vacated in the upper storey for myself and my men. I had noticed while talking to him that the end room on the ground floor was untenanted, so I told him I would stay in it and that he need only have one room vacated in the upper storey for my men. The room I had elected to spend the night in had no door, but this did not matter for I had been told that the last kill had taken place on the far side of the river and I knew the man-eater would not attempt to cross the river while it was in flood.

The room had no furniture of any kind, and after my men had swept all the straw and bits of rags out of it, complaining as they did so that the last tenant must have been a very dirty person, they spread my groundsheet on the mud floor and made up my bed. I ate my dinner—which my servant cooked on an open fire in the courtyard—sitting on my bed, and as I had done a lot of walking during the twelve hours I had been on my feet it did not take me long to get to sleep. The sun was just rising next morning, flooding the room with light, when on hearing a slight sound in the room I opened my eyes and saw a man sitting on the floor near my bed. He was about fifty years of age, and *in the last stage of leprosy*. On seeing that I was awake this unfortunate living death said he hoped I had spent a comfortable night in his room. He went on to say that he had been on a two-days' visit to friends in an adjoining village, and finding me asleep in his room on his return had sat near my bed and waited for me to awake.

Leprosy, the most terrible and the most contagious of all diseases in the East, is very prevalent throughout Kumaon, and especially bad in the Almora district. Being fatalists the people look upon the disease as a visitation from God, and neither segregate the afflicted nor take any precautions against infection. So, quite evidently, the headman did not think it necessary to warn me that the room I had selected to stay in had for years been the home of a leper. It did not take me long to dress that morning, and as soon as our guide was ready we left the village.

Moving about as I have done in Kumaon I have always gone in dread of leprosy, and I have never felt as unclean as I did after my night in that poor unfortunate's room. At the first stream we came to I called a halt, for my servant to get breakfast ready for me and for my men to have their food. Then, telling my men to wash my groundsheet and lay my bedding out in the sun, I took a bar of carbolic soap and went down the stream to where there was a little pool surrounded by great slabs of rock. Taking off every stitch of clothing I had worn in that room, I washed it all in the pool and, after laying it out on the slabs of rock, I used the remainder of the soap to scrub myself as I had never scrubbed myself before. Two hours later, in garments that had shrunk a little from the rough treatment they had received, I returned to my men feeling clean once again, and with a hunter's appetite for breakfast.

Our guide was a man about four foot six inches tall with a

F

big head crowned with a mop of long hair; a great barrel of a
body, short legs, and few words. When I asked him if we
had any stiff climbing to do, he stretched out his open hand,
and answered, 'Flat as that.' Having said this he led us down
a very steep hill into a deep valley. Here I expected him to
turn and follow the valley down to its junction with the river.
But no. Without saying a word or even turning his head he
crossed the open ground and went straight up the hill on the
far side. This hill, in addition to being very steep and overgrown with thorn bushes, had loose gravel on it which made
the going difficult, and as the sun was now overhead and very
hot, we reached the top in a bath of sweat. Our guide, whose
legs appeared to have been made for climbing hills, had not
turned a hair.

There was an extensive view from the top of the hill, and
when our guide informed us that we still had the two high
hills in the foreground to climb before reaching the Panar
river Panwa, the mason, who was carrying a bundle containing presents for his family and a greatcoat made of heavy dark
material, handed the coat to the guide and said that as he was
making us climb all the hills in Kumaon he could carry the
coat for the rest of the way. Unwinding a length of goathair
cord from round his body the guide folded up the coat and
strapped it securely to his back. Down and up we went and
down and up again, and then away down in a deep valley we
saw the river. So far we had been going over trackless
ground, without a village in sight, but we now came on a
narrow path running straight down to the river. The nearer
we got to the river the less I liked the look of it. The path
leading to the water and up the far side showed that there was
a ford here, but the river was in flood and the crossing
appeared to me to be a very hazardous one. The guide
assured us, however, that it was perfectly safe to cross, so
removing my shoes and stockings I linked arms with Panwa
and stepped into the water. The river was about forty yards
wide and from its broken surface I judged it was running

over a very rough bed. In this I was right, and after stubbing
my toes a few times and narrowly avoiding being washed off
our feet we struggled out on the far bank.

Our guide had followed us into the river and, on looking
back, I saw that the little man was in difficulties. The water
which for us had been thigh deep was for him waist deep and
on reaching the main stream, instead of bracing his back
against it and walking crab fashion, he very foolishly faced up
stream with the result that he was swept over backwards and
submerged under the fast-running current. I was barefoot
and helpless on the sharp stones, but Panwa—to whom sharp
stones were no obstacle—threw down the bundle he was
carrying and without a moment's hesitation sprinted along
the bank to where, fifty yards farther down, a big slab of rock
jutted into the river at the head of a terrifying rapid. Running
out on to this wet and slippery rock Panwa lay down, and as
the drowning man was swept past, grabbed him by his long
hair and after a desperate struggle drew him on to the rock.
When the two men rejoined me—the guide looking like a
drowned rat—I complimented Panwa on his noble and brave
act in having saved the little man's life, at great risk to his
own. After looking at me in some surprise Panwa said, 'Oh,
it was not his life that I wanted to save, but my new coat that
was strapped to his back.' Anyway, whatever the motive, a
tragedy had been averted, and after my men had linked arms
and crossed safely I decided to call it a day and spend the
night on the river bank. Panwa, whose village was five miles
farther up the river, now left me, taking with him the guide,
who was frightened to attempt a second crossing of the river.

3

Next morning we set out to find Sanouli, where the last
human kill had taken place. Late in the evening of that
day we found ourselves in a wide open valley, and as there
were no human habitations in sight, we decided to spend the
night on the open ground. We were now in the heart of the

man-eater's country and after a very unrestful night, spent on cold wet ground, arrived about midday at Sanouli. The inhabitants of this small village were overjoyed to see us and they very gladly put a room at the disposal of my men, and gave me the use of an open platform with a thatched roof.

The village was built on the side of a hill overlooking a valley in which there were terraced fields, from which a paddy crop had recently been harvested. The hill on the far side of the valley sloped up gradually, and a hundred yards from the cultivated land there was a dense patch of brushwood, some twenty acres in extent. On the brow of the hill, above this patch of brushwood, there was a village, and on the shoulder of the hill to the right another village. To the left of the terraced fields the valley was closed in by a steep grassy hill. So, in effect, the patch of brushwood was surrounded on three sides by cultivated land, and on the fourth by open grass land.

While breakfast was being got ready, the men of the village sat round me and talked. During the second half of March and the first half of April, four human beings had been killed in this area by the man-eater. The first kill had taken place in the village on the shoulder of the hill, the second and third in the village on the brow of the hill, and the fourth in Sanouli. All four victims had been killed at night and carried some five hundred yards into the patch of brushwood, where the leopard had eaten them at his leisure, for—having no fire-arms—the inhabitants of the three villages were too frightened to make any attempt to recover the bodies. The last kill had taken place six days before, and my informants were convinced that the leopard was still in the patch of brushwood.

I had purchased two young male goats in a village we passed through earlier that day, and towards evening I took the smaller one and tied it at the edge of the path of brushwood to test the villagers' assertion that the leopard was still in the cover. I did not sit over the goat, because there were

no suitable trees near by and also because clouds were banking up and it looked as though there might be rain during the night. The platform that had been placed at my disposal was open all round, so I tied the second goat near it in the hope that if the leopard visited the village during the night it would prefer a tender goat to a tough human being. Long into the night I listened to the two goats calling to each other. This convinced me that the leopard was not within hearing distance. However, there was no reason why he should not return to the locality, so I went to sleep hoping for the best.

There was a light shower during the night and when the sun rose in a cloudless sky every leaf and blade of grass was sparkling with raindrops and every bird that had a song to sing was singing a joyful welcome to the day. The goat near my platform was contentedly browsing off a bush and bleating occasionally, while the one across the valley was silent. Telling my servant to keep my breakfast warm, I crossed the valley and went to the spot where I had tied up the smaller goat. Here I found that, some time before the rain came on, a leopard had killed the goat, broken the rope, and carried away the kill. The rain had washed out the drag-mark, but this did not matter for there was only one place to which the leopard could have taken his kill, and that was into the dense patch of brushwood.

Stalking a leopard, or a tiger, on its kill is one of the most interesting forms of sport I know of, but it can only be indulged in with any hope of success when the conditions are favourable. Here the conditions were not favourable, for the brushwood was too dense to permit of a noiseless approach. Returning to the village, I had breakfast and then called the villagers together, as I wanted to consult them about the surrounding country. It was necessary to visit the kill to see if the leopard had left sufficient for me to sit over and, while doing so, I would not be able to avoid disturbing the leopard. What I wanted to learn from the villagers was whether there was any other heavy cover, within a reasonable distance, to which the leopard could retire on being disturbed by me. I was told that there was no such cover nearer than two miles, and that to get to it the leopard would have to cross a wide stretch of cultivated land.

At midday I returned to the patch of brushwood and, a hundred yards from where he had killed it, I found all that the leopard had left of the goat—its hooves, horns, and part of its stomach. As there was no fear of the leopard leaving the cover at that time of day for the jungle two miles away, I tried for several hours to stalk it, helped by bulbuls, drongos, thrushes, and scimetar-babblers, all of whom kept me informed of the leopard's every movement. In case any should question why I did not collect the men of the three villages and get them to drive the leopard out on to the open ground, where I could have shot it, it is necessary to say that this could not have been attempted without grave danger to the beaters. As soon as the leopard found he was being driven towards open ground, he would have broken back and attacked anyone who got in his way.

On my return to the village after my unsuccessful attempt to get a shot at the leopard, I went down with a bad attack of malaria and for the next twenty-four hours I lay on the platform in a stupor. By the evening of the following day the fever had left me and I was able to continue the hunt. On their own initiative the previous night my men had tied out the second goat where the first had been killed, but the leopard had not touched it. This was all to the good, for the leopard would now be hungry, and I set out on that third evening full of hope.

On the near side of the patch of brushwood, and about a hundred yards from where the goat had been killed two nights previously, there was an old oak tree. This tree was growing out of a six-foot-high bank between two terraced fields and was leaning away from the hill at an angle that made it possible for me to walk up the trunk in my rubber-soled shoes. On the underside of the trunk and about fifteen feet from the ground there was a branch jutting out over the lower field. This branch, which was about a foot thick, offered a very uncomfortable and a very unsafe seat for it was hollow and rotten. However, as it was the only branch on the tree, and as there were no other trees within a radius of several hundred yards, I decided to risk the branch.

As I had every reason to believe—from the similarity of the pug-marks I had found in the brushwood to those I had seen in April on the path leading to the homestead where the girl was killed—that the leopard I was dealing with was the Panar man-eater, I made my men cut a number of long blackthorn shoots. After I had taken my seat with my back to the tree and my legs stretched out along the branch, I made the men tie the shoots into bundles and lay them on the trunk of the tree and lash them to it securely with strong rope. To the efficient carrying out of these small details I am convinced I owe my life.

Several of the blackthorn shoots, which were from ten to twenty feet long, projected on either side of the tree; and as I had nothing to hold on to, to maintain my balance, I gathered the shoots on either side of me and held them firmly pressed between my arms and my body. By five o'clock my preparations were complete. I was firmly seated on the branch with my coat collar pulled up well in front to protect my throat, and my soft hat pulled down well behind to protect the back of my neck. The goat was tied to a stake driven into the field thirty yards in front of me, and my men were sitting out in the field smoking and talking loudly.

Up to this point all had been quiet in the patch of brushwood, but now, a scimetar-babbler gave its piercing alarm call followed a minute or two later by the chattering of several white-throated laughing thrushes. These two species of birds are the most reliable informants in the hills, and on hearing them I signalled to my men to return to the village. This they appeared to be very glad to do, and as they walked away, still talking loudly, the goat started bleating. Nothing happened for the next half-hour and then, as the sun was fading off the hill above the village, two drongos that had been sitting on the tree above me, flew off and started to bait some animal on the open ground between me and the patch of brushwood. The goat while calling had been facing in the direction of the village, and it now turned round, facing me, and stopped calling. By watching the goat I could follow the movements of the animal that the drongos were baiting and that the goat was interested in, and this animal could only be a leopard.

The moon was in her third quarter and there would be several hours of darkness. In anticipation of the leopard's coming when light conditions were not favourable, I had armed myself with a twelve-bore double-barrelled shot gun loaded with slugs, for there was a better chance of my hitting the leopard with eight slugs than with a single rifle bullet. Aids to night shooting, in the way of electric lights and torches, were not used in India at the time I am writing about, and all that one had

to rely on for accuracy of aim was a strip of white cloth tied
round the muzzle of the weapon.

Again nothing happened for many minutes, and then I felt
a gentle pull on the blackthorn shoots I was holding and
blessed my forethought in having had the shoots tied to the
leaning tree, for I could not turn round to defend myself and
at best the collar of my coat and my hat were poor protection.
No question now that I was dealing with a man-eater, and a
very determined man-eater at that. Finding that he could
not climb over the thorns, the leopard, after his initial pull,
had now got the butt ends of the shoots between his teeth
and was jerking them violently, pulling me hard against the
trunk of the tree. And now the last of the daylight faded out
of the sky and the leopard, who did all his human killing in
the dark, was in his element and I was out of mine, for in the
dark a human being is the most helpless of all animals and—
speaking for myself—his courage is at its lowest ebb. Having
killed four hundred human beings at night, the leopard was
quite unafraid of me, as was evident from the fact that while
tugging at the shoots, he was growling loud enough to be
heard by the men anxiously listening in the village. While
this growling terrified the men, as they told me later, it had
the opposite effect on me, for it let me know where the
leopard was and what he was doing. It was when he was
silent that I was most terrified, for I did not know what his
next move would be. Several times he had nearly unseated
me by pulling on the shoots vigorously and then suddenly
letting them go, and now that it was dark and I had nothing
stable to hold on to I felt sure that if he sprang up he would
only need to touch me to send me crashing to the ground.

After one of these nerve-racking periods of silence the
leopard jumped down off the high bank and dashed towards
the goat. In the hope that the leopard would come while
there was still sufficient light to shoot by, I had tied the goat
thirty yards from the tree to give me time to kill the leopard
before it got to the goat. But now, in the dark, I could not

save the goat—which, being white, I could only just see as an indistinct blur—so I waited until it had stopped struggling and then aimed where I thought the leopard would be and pressed the trigger. My shot was greeted with an angry grunt and I saw a white flash as the leopard went over backwards, and disappeared down another high bank into the field beyond.

For ten or fifteen minutes I listened anxiously for further sounds from the leopard, and then my men called out and asked if they should come to me. It was now quite safe for them to do so, provided they kept to the high ground. So I told them to light pine torches, and thereafter carry out my instructions. These torches, made of twelve to eighteen inches long splinters of resin-impregnated pine-wood cut from a living tree, give a brilliant light and provide the remote villages in Kumaon with the only illumination they have ever known.

After a lot of shouting and running about, some twenty men each carrying a torch left the village and, following my instructions, circled round above the terraced fields and approached my tree from behind. The knots in the ropes securing the blackthorn shoots to the tree had been pulled so tight by the leopard that they had to be cut. After the thorns had been removed men climbed the tree and helped me down, for the uncomfortable seat had given me cramp in my legs.

The combined light from the torches lit up the field on which the dead goat was lying, but the terraced field beyond was in shadow. When cigarettes had been handed round I told the men I had wounded the leopard but did not know how badly, and that we would return to the village now and I would look for the wounded animal in the morning. At this, great disappointment was expressed. 'If you have wounded the leopard it must surely be dead by now.' 'There are many of us, and you have a gun, so there is no danger.' 'At least let us go as far as the edge of the field and see if the leopard has left a blood trail.' After all arguments for and against going to look for the leopard immediately had been exhausted I consented against my better judgement to go as far as the edge of the field, from where we could look down on the terraced field below.

Having acceded to their request, I made the men promise that they would walk in line behind me, hold their torches high, and not run away and leave me in the dark if the leopard charged. This promise they very willingly gave, and after the torches had been replenished and were burning brightly we set off, I walking in front and the men following five yards behind.

Thirty yards to the goat, and another twenty yards to the edge of the field. Very slowly, and in silence, we moved forward. When we reached the goat—no time now to look for a blood trail—the farther end of the lower field came into view. The nearer we approached the edge, the more of this field became visible, and then, when only a narrow strip remained in shadow from the torches, the leopard, with a succession of angry grunts, sprang up the bank and into full view.

There is something very terrifying in the angry grunt of a charging leopard, and I have seen a line of elephants that were staunch to tiger turn and stampede from a charging leopard; so I was not surprised when my companions, all of whom were unarmed, turned as one man and bolted. Fortunately for me, in their anxiety to get away they collided with each other and some of the burning splinters of pine—held loosely in their hands—fell to the ground and continued to flicker, giving me sufficient light to put a charge of slugs into the leopard's chest.

On hearing my shot the men stopped running, and then I heard one of them say, '*Oh, no.* He won't be angry with us, for he knows that this devil has turned our courage to water.' Yes, I knew, from my recent experience on the tree, that fear of a man-eater robs a man of courage. As for running away, had I been one of the torch-bearers I would have run with the best. So there was nothing for me to be angry about. Presently, while I was making believe to examine the leopard, to ease their embarrassment, the men returned in twos and threes. When they were assembled. I asked, without looking up, 'Did you bring a bamboo pole and rope to carry the leopard back to the village?' 'Yes,' they answered eagerly, 'we left them at the foot of the tree.' 'Go and fetch them,' I said, 'for I want to get back to the village for a cup of hot tea.' The cold night-wind blowing down from the north had brought on another attack of malaria, and now that all the excitement was over I was finding it difficult to remain on my feet.

That night, for the first time in years, the people of Sanouli slept, and have continued to sleep, free from fear.

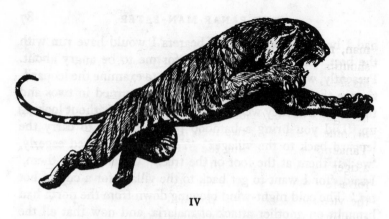

IV

The Chuka Man-Eater

I

CHUKA—which gave its name to the man-eating tiger of the Ladhya valley—is a small village of some ten ploughs on the right bank of the Sarda river near its junction with the Ladhya. From the north-west corner of the village a path runs for a quarter of a mile along a fire track before it divides, one arm going straight up a ridge to Thak village and the other diagonally up and across the hills to Kotekindri, a village owned by the people of Chuka.

Along this latter path a man was driving two bullocks in the winter of 1936, and as he approached Chuka a tiger suddenly appeared on the fire track. With very commendable courage the man interposed himself between the tiger and his bullocks and, brandishing his stick and shouting, attempted to drive the tiger away. Taking advantage of the diversion created in their favour the bullocks promptly bolted to the village and the tiger, baulked of his prey, turned his attention to the man. Alarmed at the threatening attitude of the tiger the man turned to run and, as he did so, the tiger sprang on him. Across his shoulders the man had a heavy wooden plough, and on his back he was carrying a bag containing the provisions he needed for his stay at Chuka. While the tiger was expending its teeth and claws on the plough and bag, the

88

man, relieved of his burdens, sprinted towards the village shouting for help as he ran. His relatives and friends, hearing his shouts, rallied to his assistance and he reached the village without further incidents. One claw of the tiger had ripped his right arm from shoulder to wrist, inflicting a deep wound.

Some weeks later two men returning from the market at Tanakpur were climbing the steep path to Kotekindri, when a tiger crossed the path fifty yards ahead of them. Waiting for a few minutes to give the tiger time to move away from the vicinity of the path, the men proceeded on their way, shouting as they went. The tiger had not moved away, however, and as the leading man came abreast of it, it sprang on him. This man was carrying a sack of *gur* (unrefined sugar), half of which was on his head and the other half on his back. The tiger's teeth caught in the sack and he carried it away down the hillside, without doing the man any injury. There is no record of what the tiger thought of the captures he had made so far—a plough and a sack of *gur*—but it can be assumed he was not satisfied with his bag, for, from now on, he selected human beings who were not burdened with either ploughs or sacks.

Thak, which is about three thousand feet above Chuka, has quite a large population for a hill village. The Chand Rajas who ruled Kumaon before the advent of the Gurkhas, gave the lands of Thak to the forefathers of the present holders for their maintenance, and appointed them hereditary custodians of the Purnagiri temples. Rich lands and a considerable income from the temples have enabled the people of Thak to build themselves good substantial houses, and to acquire large herds of cattle.

On a day early in June 1937, seven men and two boys were herding the village cattle, two hundred yards to the west of Thak. At fo a.m. it was noticed that some of the cattle were beginning to stray off the open ground towards the jungle and one of the boys, aged fourteen, was sent to turn them back. Six hours later the men, who had been sleeping

through the heat of
the day, were aroused by the
barking of a *kakar* in the jungle
bordering the open ground, into which
all the cattle had by now strayed, and the
second boy who was also about fourteen was sent to drive
them out. Shortly after he entered the jungle the cattle
stampeded and as they were crossing an open ravine on their
way to the village a tiger sprang on one of the cows, and
killed it in full view of the seven men. The bellowing of the
cattle and the shouts of the men attracted the attention of the
people in the village, and a crowd soon collected on the high
ground overlooking the ravine. The mother—a widow—of
the second boy was among these people and, on hearing the
men calling to her son, she ran towards them to inquire what
had happened. On learning that her son had entered the
jungle to drive out the cattle and had not returned, she set
off to look for him. At this moment the parents of the first
boy arrived on the scene and it was only when they asked
where their son was that the seven men remembered they
had not seen him since 10 a.m.

Followed by the large crowd of men who had now collected
in the ravine near the dead cow, the distracted mother went
into the jungle and found her son where the tiger had killed
and left him, and under a near-by bush the parents of the first
boy found their son dead and partly eaten. Close to this boy
was a dead calf. From the accounts the villagers subsequently
gave me of the tragic happenings of that day, I believe that
the tiger was lying up in the jungle overlooking the ground

on which the cattle were grazing, and when the calf, unseen by the men, entered the jungle the tiger killed it, and before it was able to carry it away the boy either inadvertently or through curiosity approached the calf, was killed, dragged under the bush, and partly eaten. After this the tiger apparently lay up near his two kills until 4 p.m. when a *kakar* on its way to drink at the small pool on the edge of the clearing either saw or smelt it and started barking. This aroused the men to the fact that the cattle had strayed into the jungle, and the second boy who was sent to drive them out had the ill luck to go straight to the spot where the tiger was guarding his kills.

The killing of the second boy was evidently witnessed by the cattle, who rallied to his rescue—I have seen this happen with both cows and buffaloes—and after driving the tiger from the boy they stampeded. Enraged at being driven off his kills, and at the rough treatment he had quite possibly received in the process, the tiger followed the stampeding cattle and wreaked his vengeance on the first one he was able to get hold of. Had the herd not run right on into the village he would probably not have been satisfied with killing only one of his attackers. In a similar case of attempted rescue I once saw an entire herd of five buffaloes wiped out in a titanic fight with an enraged tiger. The tiger killed one of their number and the other four big-hearted animals attacked him and fought on until the last of them had been killed. The tiger evidently suffered severely in the fight, for when he left the battle-ground he left a trail of blood.

The seemingly wanton slaughter of two human beings and two animals on the same day—resulting I am convinced from the tiger's having been disturbed on his first kill—caused a great outcry in the districts of Naini Tal and Almora, and every effort was made to kill the tiger. On several occasions district officials sat up all night on *machans* over kills, and though the tiger had been wounded on two occasions—unfortunately only with buckshot—he continued to prey on human beings, and claimed yet another victim from the ill-fated village of Thak.

Two hundred yards above Thak there is a wheat field. The crop had been cut from this field and two boys were grazing a few cattle on the stubble. For safety's sake the boys, who were brothers and orphans ten and twelve years of age, were sitting in the middle of the field. On the far side of the field, from the village, there was a light fringe of bushes. From these the hill went steeply up for a thousand feet, and from anywhere on the hill the two boys sitting in the open would have been visible. Towards the afternoon a cow strayed towards the bushes and the boys, keeping close together, set off to drive it back on to the field. The elder boy was leading and as he passed a bush the tiger, who was lying in wait, pounced on him and carried him away. The younger boy fled back to the village and dashing up to a group of men fell sobbing at their feet. When the boy was able to speak coherently he told the men that a big red animal—it was the first tiger he had ever seen—had carried away his brother. A search party was hastily organized and with very commendable bravery the blood trail was followed for about a mile into the densely wooded Suwar Gadh ravine to the east of the village. Night was now closing in so the party returned to Thak. The following day, assisted by men from adjoining villages, a day-long search was made but all that was found of the boy was his red cap and his torn and bloodstained clothes. That was the Chuka man-eater's last human victim.

I do not think it is possible to appreciate courage until the danger that brought it into being has been experienced. Those who have never lived in an area in which a man-eating tiger is operating may be inclined to think that there was nothing courageous in a mother going to look for her son, in two boys grazing cattle, or in a party of men going out to look for a missing boy. But to one who has lived in such an area the entry of the mother into a dense patch of jungle in which she knew there was an angry tiger; the two small boys

sitting close together for protection; and the party of unarmed men following on the blood trail left by a man-eater,
were acts calling for a measure of courage that merited the
greatest admiration.

2

The Chuka man-eater was now disorganizing life for everyone in the Ladhya valley, and shortly after Ibbotson had been
appointed Deputy Commissioner-in-Charge of the three
districts of Naini Tal, Almora, and Garhwal, we joined forces
to try to rid his division of the menace.

It was early afternoon on a sweltering hot day in April 1937
that Ibby, his wife Jean, and I, alighted from our motor bus
at the Boom above Baramdeo. We had left Naini Tal in the
early hours of the morning and, travelling via Haldwani and
Tanakpur, arrived at the Boom at the hottest time of the day,
covered in dust from head to foot, and with many sore spots
in unseen and tender places. A cup of tea drunk while sitting
on yielding sand on the bank of the Sarda river helped to
restore our spirits, and taking the short cut along the river
bank we set off on foot to Thuli Gadh where our tents, sent
in advance, had been pitched.

Starting after breakfast next morning we went to Kaladhunga. The distance between Thuli Gadh and Kaladhunga
via the Sarda gorge is eight miles, and via Purnagiri, fourteen
miles. The Ibbotsons and I went through the gorge while
our servants and the men carrying our kit went via Purnagiri.
The gorge is four miles long and was at one time traversed by
a tramway line blasted out of the rock cliff by J. V. Collier
when extracting the million cubic feet of *sal* timber presented
by the Nepal Durbar to the Government of India as a thank-
offering after the first world war. The tramway line has long
since been swept away by landslides and floods, and these
four miles necessitate a great deal of rock climbing where a
false step or the slipping of a hand-hold would inevitably
precipitate one into the cold river. We negotiated the gorge

without mishap, and at the upper end, where Collier's tram-line entered the forest, we caught two fish in a run where a rock the size of a house juts out into the river.

Word had been sent ahead for the *patwaris* and forest guards working in the area to meet us at Kaladhunga and give us the latest news of the man-eater. We found four men awaiting our arrival at the bungalow and the reports they gave us were encouraging. No human beings had been killed within the past few days, and the tiger was known to be in the vicinity of Thak village where three days previously it had killed a calf.

Kaladhunga is a gently rising cone-shaped peninsula roughly four miles long and a mile wide, surrounded on three sides by the Sarda river and backed on the fourth by a ridge of hills five thousand feet high. The bungalow, a three-roomed house with a wide veranda, faces east and is situated at the northern or upper end of the peninsula. The view from the veranda as the morning sun rises over the distant hills and the mist is lifting is one of the most pleasing prospects it is possible to imagine. Straight in front, and across the Sarda, is a wide open valley running deep into Nepal. The hills on either side are densely wooded, and winding through the valley is a river fringed with emerald-green elephant grass. As far as the eye can see there are no human habitations and, judging from the tiger and other animal calls that can be heard from the bungalow, there appears to be an abundant stock of game in the valley. It was from this valley that Collier extracted the million cubic feet of timber.

We spent a day at Kaladhunga, and while our men went on to

Chuka to pitch our tents and make camp we fished, or, to be correct, the Ibbotsons fished while I, who had been laid up with malaria the previous night, sat on the bank and watched. From the broken water below the bungalow to the point of the peninsula—a stretch of some five hundred yards —the Ibbotsons, who are expert thread-line casters, combed the river with their one-inch spoons without moving a fish. The small river that flows down the Nepal valley joins the Sarda opposite the point of the peninsula. Here the Sarda widens out and shallows, and flows for two hundred yards over rocks before entering a big pool. It was at the upper end of this run and well out in the river that Ibby hooked his first fish—an eight-pounder—which needed careful handling on the light tackle before it was eventually brought to the bank and landed.

All keen anglers delight in watching others of the craft indulging in this, one of the best of outdoor sports. For myself I would just as soon watch another fishing as fish myself, especially when fish are on the take, the foothold uncertain—as it always is in the Sarda—and the river fairly fast. Shortly after Ibby killed his fish, Jean—who was fishing in broken water thirty yards from the bank—hooked a fish. Her reel only held a hundred yards of thread line, and fearing that the fish would make for the pool and break her, she attempted to walk backwards while playing the fish, and in doing so lost her footing and for a long minute all that was visible of her was the toe of one foot and the point of her rod. You will naturally assume that I, forgetting my recent attack of malaria, dashed out to her rescue. As a fact I did nothing of the kind and only sat on the bank and laughed, for to attempt to rescue either of the Ibbotsons from a watery grave would be as futile as trying to save an otter from drowning. After a long and a violent struggle Jean upended herself, and reaching the bank killed her fish, which weighed six pounds. Hardly had she done so when Ibby, in making a long cast, slipped off the rock on which he was standing and disappeared under water, rod and all.

From the bottom end of the pool below the run, the river turns to the right. On the Nepal side of this bend in the river there had stood a giant *semul* tree, in the upper branches of which a pair of ospreys had for many years built their nest. The tree had been an ideal home for the birds, for not only had it commanded an extensive view of the river, but the great branches growing out at right angles to the trunk had provided tables on which to hold and devour their slippery prey. The monsoon floods of the previous year had cut into the bank and washed away the old tree, and the ospreys had built themselves a new nest in a tall *shisham* tree standing at the edge of the forest, a hundred yards from the river.

The run was evidently the favourite fishing ground of the ospreys, and while the female sat in the nest the male kept flying backwards and forwards over the Ibbotsons' heads. Eventually tiring of this unprofitable exercise it flew farther down the river to where a few partly submerged rocks broke the surface of the water, making a small run. Fish were evidently passing this spot, and a dozen times the osprey banked steeply, closed his wings, and dropped like a plummet and, checking himself with widespread wings and tail before reaching the water, rose flapping to regain height for his next cast. At last his patience was rewarded. An unwary fish had come to the surface directly below him, and without a moment's pause he went from flat flight into a lightning dive through a hundred feet of air and plunged deep into the broken water. His needle-sharp and steel-strong talons took hold, but the catch was evidently heavier than he anticipated. Time and time again with wildly threshing wings he attempted to launch himself into the air, only to sink down again on his breast feathers. I believe he would have had to relinquish his catch had not a gust of wind blowing up river come at this critical moment to help him. As the wind reached him he turned downstream and, making one last desperate effort, got the fish clear of the water. Home was in the opposite direction from that in which he was now heading but to turn was impossible, so, selecting a great slab of rock on the bank on which to land, he made straight for it.

I was not the only one who had been watching the osprey, for he had hardly landed on the rock when a woman who had been washing clothes on the Nepal side of the river called out excitedly, and a boy appeared on the high bank above her. Running down the steep track to where the woman was standing, he received his instructions and set off along the boulder-strewn bank at a pace that threatened his neck and limbs at every stride. The osprey made no attempt to carry off his prey, and as the boy reached the rock it took to the air, circling round his head as he held up the fish—which appeared to be about four pounds in weight—for the woman to see.

For some time thereafter I lost sight of the osprey, and we had finished our lunch before I again saw him quartering the air above the run in which he had caught the fish of which the boy had deprived him. Back and forth he flew for many minutes, always at the same height, and then he banked, dropped fifty feet, again banked and then plunged straight into the water. This time his catch was lighter—a *kalabas* about two pounds in weight—and without effort he lifted it clear of the water and, holding it like a torpedo to reduce wind pressure, made for his nest. His luck was out that day, however, for he had only covered half the distance he had to go, when a Pallas fish-eagle twice his weight and size came up from behind, rapidly overtaking him. The osprey saw him coming and altering his course a point to the right headed for the forest with the object of shaking off his pursuer among the branches of the trees. The eagle realizing the object of this manoeuvre emitted a scream of rage and increased his speed. Only twenty yards more to go to reach safety, but the risk was too great to take and, relinquishing his hold of the *kalabas*, the osprey—only just in time—hurled himself straight into the air. The fish had not fallen a yard before the eagle caught it and, turning in a graceful sweep,

made off down river in the direction from which he had come. He was not to escape with his booty as easily as he expected, however, for he had only gone a short distance on his return journey when the pair of crows that fed on the osprey's leavings set off to bait him, and to shake them off he too was compelled to take to the forest. At the edge of the forest the crows turned back and the eagle had hardly disappeared from view when falling out of the sky came two Tawny eagles going at an incredible speed in the direct line the Pallas eagle had taken. I very greatly regret I did not see the end of that chase for, from the fact that while I watched neither of the birds rose out of the forest, I suspect that the Pallas eagle retained his hold on the fish too long. I have only once seen a more interesting chase. On that occasion I was taking a line of eighteen elephants through grass and the ten guns and five spectators who were sitting on the elephants, shooting black partridge, saw a bush chat pass—without once touching the ground—from a sparrowhawk that killed it just in front of our line of elephants to a red-headed merlin, then to a honey buzzard, and finally to a peregrine falcon who swallowed the little bird whole. If any of the guns or spectators who were with me that February morning read this chapter, they will recall the occurrence as having taken place on the Rudrapur Maidan.

After an early breakfast next morning we moved from Kaladhunga to Chuka, an easy march of five miles. It was one of those gorgeous days that live long in the memory of a fisherman. The sun was pleasantly warm; a cool wind blowing down from the north; a run of *chilwa* (fingerlings) in progress, and the river full of big fish madly on the take.

Fishing with light tackle we had many exciting battles, all of which we did not win. We ended the day, however, with enough fish to feed our camp of thirty men.

3

To assist us in our campaign against the man-eater, and to try to prevent further loss of human life, six young male buffaloes had been sent up from Tanakpur in advance of us, to be used as bait for the tiger. On our arrival at Chuka we were told that the buffaloes had been tied out for three nights, and that though a tiger's pug-marks had been seen near several of them, none had been killed. During the next four days we visited the buffaloes in the early morning; tried to get in touch with the tiger during the day, and in the evening accompanied the men engaged in tying out the buffaloes. On the fifth morning we found that a buffalo we had tied up at Thak, at the edge of the jungle in which the two boys had lost their lives, had been killed and carried off by a tiger. Instead of taking its kill into the dense jungle as we had expected, the tiger had taken it across an open patch of ground, and up on to a rocky knoll. This it had evidently done to avoid passing near a *machan* from which it had been fired at—and quite possibly wounded—on two previous occasions. After the buffalo had been dragged for a short distance its horns got jammed between two rocks, and being unable to free it, the tiger had eaten a few pounds of flesh from the hindquarters of the kill and then left it. In casting round to see in which direction the tiger had gone, we found its pug-marks in a buffalo wallow, between the kill and the jungle. These pug-marks showed that the killer of the buffalo was a big male tiger.

It was generally believed by the District Officials—on what authority I do not know—that the man-eater was a tigress. On showing them the tracks in the buffalo wallow we were told by the villagers that they could not distinguish between the pug-marks of different tigers and that they did not know

whether the man-eater was male or female, but that they did
know it had a broken tooth. In all the kills, human as well as
animal, that had taken place near their village they had
noticed that one of the tiger's teeth only bruised the skin and
did not penetrate it. From this they concluded that one of
the man-eater's canine teeth was broken.

Twenty yards from the kill there was a *jamun* tree. After
we had dragged the kill out from between the rocks we
sent a man up the tree to break a few twigs that were
obstructing a view of the kill from the only branch of the
tree on which it was possible to sit. This isolated tree on
the top of the knoll was in full view of the surrounding
jungle, and though the man climbed it and broke the twigs
with the utmost care, I am inclined to think he was seen by
the tiger.

It was now 11 a.m., so, sending our men back to the
village to have their midday meal, Ibby and I selected a bush
under which to shelter from the sun and talked and dozed,
and dozed and talked throughout the heat of the day. At
2.30 p.m., while we were having a picnic lunch, some *kalege*
pheasants started chattering agitatedly at the edge of the
jungle where the buffalo had been killed, and on hearing
them our men returned from the village. While Ibby and his
big-hearted man, Sham Singh, went into the jungle where
the pheasants were chattering—to attract the tiger's attention
—I climbed silently into the *jamun* tree. Giving me a few
minutes in which to settle down, Ibby and Sham Singh came
out of the jungle and returned to our camp at Chuka, leaving
my two men at Thak.

Shortly after Ibby had gone the pheasants started chattering again and a little later a *kakar* began barking. The tiger
was evidently on the move, but there was little hope of his
crossing the open ground and coming to the kill until the sun
had set and the village had settled down for the night. The
kakar barked for a quarter of an hour or more before it
finally stopped, and from then until sunset, except for the

natural calls of a multitude of birds, the jungle—as far as the tiger was concerned—was silent.

The red glow from the setting sun had faded from the Nepal hills on the far side of the Sarda river, and the village sounds had died down, when a *kakar* barked in the direction of the buffalo wallow; the tiger was returning to his kill along the line he had taken when leaving it. A convenient branch in front of me gave a perfect rest for my rifle, and the only movement it would be necessary to make when the tiger arrived would be to lower my head on to the rifle butt. Minute succeeded minute until a hundred had been added to my age and then, two hundred yards up the hillside, a *kakar* barked in alarm and my hope of getting a shot, which I had put at ten to one, dropped to one in a thousand. It was now quite evident that the tiger had seen my man breaking the twigs off the tree, and that between sunset and the barking of this last *kakar* he had stalked the tree and seeing me on it had gone away. From then on *kakar* and *sambhar* called at intervals, each call a little farther away than the last. At midnight these alarm calls ceased altogether, and the jungle settled down to that nightly period of rest ordained by Nature, when strife ceases and the jungle folk can sleep in peace. Others who have spent nights in an Indian jungle will have noticed this period of rest, which varies a little according to the season of the year and the phases of the moon, and which as a rule extends from midnight to 4 a.m. Between these hours killers sleep, and those who go in fear of them are at peace. It may be natural for carnivora to sleep from midnight to 4 a.m., but I would prefer to think that Nature had set apart these few hours so that those who go in fear of their lives can relax and be at peace.

Day was a few minutes old when, cramped in every joint, I descended from the tree and, unearthing the thermos flask Ibby had very thoughtfully buried under a bush, indulged in a much needed cup of tea. Shortly after my two men arrived and while we were covering the kill with branches, to

protect it from vultures, the tiger called three times on a hill
half a mile away. As I passed through Thak on my way back
to camp the greybeards of the village met me and begged me
not to be discouraged by my night's failure, for, they said,
they had consulted the stars and offered up prayers and if the
tiger did not die this day it would certainly die on the next or,
may be, the day after.

A hot bath and a square meal refreshed me and at 1 p.m.
I again climbed the steep hill to Thak and was told on my
arrival that a *sambhar* had belled several times on the hill
above the village. I had set out from camp with the intention
of sitting up over a live buffalo; and, to ensure while doing
so that the tiger did not feed in one place while I was waiting
for him in another, I placed several sheets of newspaper near
the kill I had sat over the previous night. There was a well-
used cattle track through the jungle in which the villagers
said the *sambhar* had called. In a tree on the side of this track
I put up a rope seat, and to a root on the track I tied the
buffalo. I climbed into the tree at three o'clock, and an hour
later first a *kakar* and then a tiger called on the far side of the
valley a thousand yards away. The buffalo had been provided with a big feed of green grass, and throughout the
night it kept the bell I had tied round its neck ringing, but it
failed to attract the tiger. At daylight my men came for me
and they told me that *sambhar* and *kakar* had called during
the night in the deep ravine in which the boy's red cap and
torn clothes had been found, and at the lower end of which
we had tied up a buffalo at the request of the villagers.

When I got back to Chuka I found that Ibby had left camp
before dawn. News had been received late the previous
evening that a tiger had killed a bullock eight miles up the
Ladhya valley. He sat up over the kill all night without seeing
anything of the tiger, and late the following evening he
returned to camp.

4

Jean and I were having breakfast after my night in the tree
over the live buffalo, when the men engaged in tying out our
remaining five buffaloes came in to report that the one they
had tied up at the lower end of the ravine in which my men
had heard the *sambhar* and *kakar* calling the previous night
was missing. While we were being given this news Mac-
Donald, Divisional Forest Officer, who was moving camp
that day from Kaladhunga to Chuka, arrived and told us he
had seen the pug-marks of a tiger at the lower end of a ravine
where he presumed one of our buffaloes had been tied up.
These pug-marks, Mac said, were similar to those he had
seen at Thak when on a previous visit he had tried to shoot
the man-eater.

After breakfast Jean and Mac went down the river to fish
while I went off with Sham Singh to try to find out what had
become of the missing buffalo. Beyond the broken rope and
the tiger's pug-marks there was nothing to show that the
buffalo had been killed. However, on casting round I found
where one of the buffalo's horns had come in contact with the
ground and from here on there was a well-defined blood trail.
Whether the tiger lost his bearings after killing the buffalo or
whether he was trying to cover up his tracks I do not know,
for after taking the kill over most difficult ground for several
miles he brought it back to the ravine two hundred yards
from where he had started. At this point the ravine narrowed
down to a bottle-neck some ten feet wide. The tiger was
probably lying up with his kill on the far side of the narrow
neck, and as I intended sitting up for him all night I decided
to join the anglers and share their lunch before sitting up.

After fortifying the inner man I returned with Sham Singh
and three men borrowed from the fishing party, for if I found
the kill and sat up over it it would not have been safe for
Sham Singh to have gone back to camp alone. Walking well
ahead of the four men I approached the bottle-neck for the

second time, and as I did so the tiger started growling. The ravine here was steep and full of boulders and the tiger was growling from behind a dense screen of bushes, about twenty yards straight in front of me. An unseen tiger's growl at close range is the most terrifying sound in the jungle, and is a very definite warning to intruders not to approach any nearer. In that restricted space, and with the tiger holding a command-ing position, it would have been foolish to have gone any farther. So, signalling to the men to retire, and giving them a few minutes to do so, I started to walk backwards very slowly —the only safe method of getting away from any animal with which one is not anxious to make contact. As soon as I was well clear of the bottle-neck I turned and, whistling to the men to stop, rejoined them a hundred yards farther down the ravine. I now knew exactly where the tiger was, and felt confident I would be able to deal with him; so, on rejoining the men, I told them to leave me and return to the fishing party. This, however, they were very naturally frightened to do. They believed, as I did, that the tiger they had just heard growling was a man-eater and they wanted to have the protection of my rifle. To have taken them back myself would have lost me two hours, and as we were in a *sal* forest and there was not a climbable tree in sight, I had of neces-sity to keep them with me.

Climbing the steep left bank we went straight away from the ravine for two hundred yards. Here we turned to the left and after I had paced out two hundred yards we again turned to the left and came back to the ravine a hundred yards above where we had heard the tiger growling. The tables were now turned and we held the advantage of position. I knew the tiger would not go down the ravine, for he had seen human beings in that direction, only a few minutes before; nor would he go up the ravine, for in order to do so he would have to pass us. The bank on our side was thirty feet high and devoid of undergrowth, so the only way the tiger could get out of the position we had manoeuvred him into would be to go up the opposite hillside. For ten minutes we sat on the edge of the ravine scanning every foot of ground in front of us. Then, moving back a few paces, we went thirty yards to the left and again sat down on the edge and, as we did so, the man sitting next to me whispered '*Sher*', and pointed across the ravine. I could see nothing, and on asking the man how much of the tiger he could see, and to describe its position, he said he had seen its ears move and that it was near some dry leaves. A tiger's ears are not conspicuous objects at fifty yards, and as the ground was carpeted with dead leaves his description did nothing to help me locate the tiger. From the breathing of the men behind me it was evident that excitement was rising to a high pitch. Presently one of the men stood up to get a better view, and the tiger, who had been lying down facing us, got up and started to go up the hill, and as his head appeared from behind a bush I fired. My bullet, I subsequently found, went through the ruff on his neck and striking a rock splintered back, making him spring straight up into the air, and on landing he got involved with a big creeper from which he found some difficulty in freeing himself. When we saw him struggling on the ground we thought he was down for good, but when he regained his feet and galloped off Sham Singh expressed the opinion, which I shared, that he was unwounded. Leaving the men I crossed the ravine and on examining the ground found the long hairs the bullet had clipped, the splintered rock, and the torn and bitten creeper, but I found no blood.

Blood does not always flow immediately an animal has been hit, and my reconstruction of the shot may have been faulty; so it was necessary to find the kill, for it would tell me on the morrow whether or not the tiger was wounded. Here we had some difficulty, and it was not until we had gone over the ground twice that we eventually found the kill in a pool of water four feet deep, where the tiger had presumably put it to preserve it from hornets and blowflies. Sending the three men I had borrowed back to the fishing party—it was safe to

do so now—Sham Singh and I remained hidden near the
kill for an hour to listen for jungle sounds and then, hearing
none, returned to camp. After an early breakfast next morning Mac and I returned to the ravine and found that the
tiger had removed the kill from the pool, carried it a short
distance, and eaten it out leaving only the head and hooves.
This, together with the absence of blood on the ground on
which he had been lying while eating, was proof that the
tiger was not wounded and that he had recovered from his
fright.

When we got back to camp we were informed that a cow
had been killed in a wide open ravine on the far side of the
Ladhya river, and that the men who had found it had covered
it with branches. Ibby had not returned from his visit to the
village eight miles up the Ladhya, and after lunch Mac and I
went out to look at the cow. It had been covered up at midday and shortly afterwards the tiger had returned, dug it out
from under the branches, and carried it away without leaving
any mark of a drag. The forest here consisted of great big *sal*
trees without any undergrowth, and it took us an hour to find
the kill where the tiger had hidden it under a great pile of
dead leaves. In a near-by tree Mac very gallantly put up a
machan for me while I smoked and emptied his water-bottle
—the shade temperature was about a hundred and ten
degrees—and after seeing me into the tree he returned to
camp. An hour later a small stone rolling down the steep
hill on the far side of the ravine attracted my attention, and
shortly after a tigress came into view, followed by two small
cubs. This was quite evidently the first occasion on which
the cubs had ever been taken to a kill, and it was very interesting to see the pains the mother took to impress on them
the danger of the proceeding and the great caution it was
necessary to exercise. The behaviour of the cubs was as
interesting as their mother's. Step by step they followed in
her tracks; never trying to pass each other, or her; avoiding
every obstruction that she avoided no matter how small it

was, and remaining perfectly rigid when she stopped to listen, which she did every few yards. The ground was carpeted with big *sal* leaves as dry as tinder over which it was impossible to move silently; however, every pad was put down carefully and as carefully lifted, and as little sound as possible was made.

Crossing the ravine, the tigress, closely followed by the cubs, came towards me and passing behind my tree lay down on a flat piece of ground overlooking the kill, and about thirty yards from it. Her lying down was apparently intended as a signal to the cubs to go forward in the direction in which her nose was pointing, and this they proceeded to do. By what means the mother conveyed the information to her cubs that there was food for them at this spot I do not know, but that she had conveyed this information to them there was no question. Passing their mother—after she had lain down—and exercising the same caution they had been made to exercise when following her, they set out with every appearance of being on a very definite quest. I have repeatedly asserted that tigers have no sense of smell, and the cubs were providing me with ample proof of that assertion. Though the kill had only been reported to us that morning the cow had actually been killed the previous day, and before hiding it under the pile of dead leaves the tigress had eaten the greater portion of it. The weather, as I have said, was intensely hot, and it was the smell that eventually enabled Mac and me to find the kill. And here, now, were two hungry cubs ranging up and down, back and forth, passing and repassing a dozen times within a yard of the kill and yet not being able to find it. It was the blowflies that disclosed its position and at length enabled them to find it. Dragging it out from under the leaves the cubs sat down together to have their meal. The tigress had watched her cubs as intently as I had and only once, when they were questing too far afield, had she spoken to them. As soon as the kill had been found the mother turned on to her back with her legs in the air and went to sleep.

H

As I watched the cubs feeding I was reminded of a scene I had witnessed some years previously at the foot of Trisul. I was lying on a ridge scanning with field glasses a rock cliff opposite me for *thar*, the most sure-footed of all Himalayan goats. On a ledge halfway up the cliff a *thar* and her kid were lying asleep. Presently the *thar* got to her feet, stretched herself, and the kid immediately started to nuzzle her and feed. After a minute or so the mother freed herself, took a few steps along the ledge, poised for a moment, and then jumped down on to another and a narrower ledge some twelve to fifteen feet below her. As soon as it was left alone the kid started running backwards and forwards, stopping every now and then to peer down at its mother, but unable to summon the courage to jump down to her for, below the few-inches-wide ledge, was a sheer drop of a thousand feet. I was too far away to hear whether the mother was encouraging her young, but from the way her head was turned I believe she was doing so. The kid was now getting more and more agitated and, possibly fearing that it would do something foolish, the mother went to what looked like a mere crack in the vertical rock face and, climbing it, rejoined her young. Immediately on doing so she lay down, presumably to prevent the kid from feeding. After a little while she again got to her feet, allowed the kid to drink for a minute, poised carefully on the brink, and jumped down, while the kid again ran backwards and forwards above her. *Seven*

times in the course of the next half-hour this procedure was gone through, until finally the kid, abandoning itself to its fate, jumped, and landing safely beside its mother was rewarded by being allowed to drink its fill. The lesson, to teach her young that it was safe to follow where she led, was over for that day. Instinct helps, but it is the infinite patience of the mother and the unquestioning obedience of her offspring that enable the young of all animals in the wild to grow to maturity. I regret I lacked the means, when I had the opportunity, of making cinematograph records of the different species of animals I have watched training their young, for there is nothing more interesting to be seen in a jungle.

When the cubs finished their meal they returned to their mother and she proceeded to clean them, rolling them over and licking off the blood they had acquired while feeding. When this job was finished to her entire satisfaction she set off, with the cubs following close behind, in the direction of a shallow ford in the Ladhya, for nothing remained of the kill and there was no suitable cover for her cubs on this side of the river.

I did not know, and it would have made no difference if I had, that the tigress I watched with such interest that day would later, owing to gunshot wounds, become a man-eater and a terror to all who lived or worked in the Ladhya valley and the surrounding villages.

5

The kill at Thak, over which I had sat the first night, had been uncovered to let the vultures eat it, and another buffalo had been tied up at the head of the valley to the west of the village and about two hundred yards from the old kill. Four mornings later the headman of Thak sent word to us that this buffalo had been killed by a tiger and carried away.

Our preparations were soon made, and after a terribly hot climb Ibby and I reached the scene of the kill at about mid-day. The tiger, after killing the buffalo and breaking a very strong rope, had picked up the kill and gone straight down into the valley. Telling the two men we had brought to carry our lunch to keep close behind us, we set off to follow the drag. It soon became apparent that the tiger was making for some definite spot, for he led us for two miles through dense undergrowth, down steep banks, through beds of nettles and raspberry bushes, over and under fallen trees, and over great masses of rock until finally he deposited the kill in a small hollow under a box tree shaped like an umbrella. The buffalo had been killed the previous night and the fact that the tiger had left it without having a meal was disquieting. However, this was to a great extent offset by the pains he had taken in getting the kill to this spot, and if all went well there was every reason to hope that he would return to his kill, for, from the teeth-marks on the buffalo's neck, we knew he was the man-eater we were looking for and not just an ordinary tiger.

Our hot walk up to Thak and subsequent descent down the densely wooded hillside, over difficult ground, had left us in a bath of sweat, and while we rested in the hollow having lunch and drinking quantities of tea, I cast my eyes round for a convenient tree on which to sit and, if necessary, in which to pass the night. Growing on the outer edge of the hollow and leaning away from the hill at an angle of forty-five degrees was a *ficus* tree. This, starting life in some decayed part of a giant of the forest, had killed the parent tree by weaving a trellis round it, and this trellis was now in course of coalescing to form a trunk for the parasite. Ten feet from the ground, and where the trellis had stopped and the parent tree had rotted and fallen away, there appeared to be a comfortable seat on which I decided to sit.

Lunch eaten and a cigarette smoked, Ibby took our two men sixty yards to the right and sent them up a tree to shake the branches and pretend they were putting up a *machan*, to distract the tiger's attention in case he was lying up close by and watching us, while I as silently as possible climbed into the *ficus* tree. The seat I had selected sloped forward and was cushioned with rotten wood and dead leaves and, fearing that if I brushed them off the sound and movement might be detected by the tiger, I left them as they were and sat down

on them, hoping devoutly that there were no snakes in the
hollow trunk below me or scorpions in the dead leaves.
Placing my feet in an opening in the trellis, to keep from
slipping forward, I made myself as comfortable as conditions
permitted, and when I had done so Ibby called the men off
the tree and went away talking to them in a loud voice.

The tree in which I had elected to sit was, as I have already
said, leaning outwards at an angle of forty-five degrees, and
ten feet immediately below me there was a flat bit of ground
about ten feet wide and twenty feet long. From this flat
piece of ground the hill fell steeply away and was overgrown
with tall grass and dense patches of brushwood, beyond
which I could hear a stream running. An ideal place for a
tiger to lie up in.

Ibby and the two men had been gone about fifteen minutes
when a red monkey on the far side of the valley started barking to warn the jungle folk of the presence of a tiger. From
the fact that this monkey had not called when we were
coming down the hill, following the drag, it was evident that
the tiger had not moved off at our approach and that he was
now coming to investigate—as tigers will—the sounds he had
heard in the vicinity of his kill. Monkeys are blessed with
exceptionally good eyesight, and though the one that was
calling was a quarter of a mile away, it was quite possible that
the tiger he was calling at was close to me.

I was sitting facing the hill with the kill to my left front,
and the monkey had only called eight times when I heard
a dry stick snap down the steep hillside behind me. Turning
my head to the right and looking through the trellis, which
on this side extended a little above my head, I saw the tiger
standing and looking in the direction of my tree, from a
distance of about forty yards. For several minutes he stood
looking alternately in my direction and then in the direction
of the tree the two men had climbed, until eventually,
deciding to come in my direction, he started up the steep
hillside. It would not have been possible for a human being

to have got over that steep and difficult ground without using his hands and without making considerable noise, but the tiger accomplished the feat without making a sound. The nearer he came to the flat ground the more cautious he became and the closer he kept his belly to the ground. When he was near the top of the bank he very slowly raised his head, took a long look at the tree the men had climbed, and satisfied that it was not tenanted sprang up on to the flat ground and passed out of sight under me. I expected him to reappear on my left and go towards the kill, and while I was waiting for him to do so I heard the dry leaves under the tree being crushed as he lay down on them.

For the next quarter of an hour I sat perfectly still, and as no further sounds came to me from the tiger I turned my head to the right, and craning my neck looked through an opening in the trellis, and saw the tiger's head. If I had been able to squeeze a tear out of my eye and direct it through the opening it would, I believe, have landed plumb on his nose. His chin was resting on the ground and his eyes were closed. Presently he opened them, blinked a few times to drive away the flies, then closed them again and went to sleep. Regaining my position I now turned my head to the left. On this side there was no trellis nor were there any branches against which I could brace myself, and when I had craned my neck as far as I could without losing my balance I looked down and found I could see most of the tiger's tail, and a part of one hind leg.

The situation needed consideration. The bole of the tree against which I had my back was roughly three feet thick and afforded ideal cover, so there was no possibility of the tiger seeing me. That he would go to the kill if not disturbed was certain and the question was, when would he go? It was a hot afternoon, but the spot he had selected to lie on was in deep shade from my tree and, further, there was a cool breeze blowing up the valley. In these pleasant conditions he might sleep for hours and not approach the kill until day light had gone, taking with it my chance of getting a shot. The risk of waiting on the tiger's pleasure could not be taken, therefore, for apart from the reason given the time at our disposal was nearly up and this might be the last chance I would get of killing the tiger, while on that chance might depend the lives of many people. Waiting for a shot being inadvisable, then, there remained the possibility of dealing with the tiger where he lay. There were several openings in the trellis on my right through which I could have inserted the barrel of my rifle, but having done this it would not have been possible to depress the muzzle of the rifle sufficiently to get the sights to bear on the tiger's head. To have stood up, climbed the trellis, and fired over the top of it would not have been difficult. But this could not have been done without making a certain amount of noise, for the dry leaves I was sitting on would have crackled when relieved of my weight, and within ten feet of me was an animal with the keenest hearing of any in the jungle. A shot at the head end of the tiger not being feasible, there remained the tail end.

When I had both my hands on the rifle and craned my neck to the left, I had been able to see most of the tiger's tail and a portion of one hind leg. By releasing my right hand from the rifle and getting a grip of the trellis I found I could lean out far enough to see one-third of the tiger. If I could maintain this position after releasing my hold, it would be possible to disable him. The thought of disabling an animal, and a sleeping one at that, simply because he occasionally liked a change of diet was hateful. Sentiment, however, where a man-eater was concerned was out of place. I had been trying for days to shoot this tiger to save further loss of human life, and now that I had a chance of doing so the fact that I would have to break his back before killing him would not justify my throwing away that chance. So the killing would have to be done no matter how unpleasant the method might be, and the sooner it was done the better, for in bringing his kill to this spot the tiger had laid a two-mile-long scent trail, and a hungry bear finding that trail might at any moment take the decision out of my hands. Keeping my body perfectly rigid I gradually released my hold of the trellis, got both hands on the rifle, and fired a shot behind and under me which I have no desire ever to repeat. When I pressed the trigger of the 450/400 high-velocity rifle, the butt was pointing to heaven and I was looking under, not over, the sights. The recoil injured but did not break either my fingers or my wrist, as I feared it would, and as the tiger threw the upper part of his body round and started to slide down the hill on his back, I swung round on my seat and fired the second barrel into his chest. I should have felt less a murderer if, at my first shot, the tiger had stormed and raved but—being the big-hearted animal that he was—he never opened his mouth, and died at my second shot without having made a sound.

Ibby had left me with the intention of sitting up in the *jamun* tree over the buffalo which had been killed four days previously and which the vultures had for some unknown reason not eaten. He thought that if the tiger had seen me climbing into the *ficus* tree it might abandon the kill over which I was sitting and go back to its old kill at Thak and give him a shot. On hearing my two shots he came hurrying back to see if I needed his help, and I met him half a mile from the *ficus* tree. Together we returned to the scene of the killing to examine the tiger. He was a fine big male in the prime of life and in perfect condition, and would have measured—if we had had anything to measure him with—nine foot six inches between pegs, or nine foot ten over curves. And the right canine tooth in his lower jaw was broken. Later I found several pellets of buckshot embedded in different parts of his body.

The tiger was too heavy for the four of us to carry back to camp so we left him where he lay, after covering him up with grass, branches, and dead wood heaped over with big

stones, to protect him from bears. Word travelled round that night that the man-eating tiger was dead and when we carried him to the foot of the *ficus* tree next morning to skin him, more than a hundred men and boys crowded round to see him. Among the latter was the ten-year-old brother of the Chuka man-eater's last human victim.

The Talla Des Man Eater

I

NOWHERE along the foothills of the Himalayas is there a
more beautiful setting for a camp than under the Flame of
the Forest trees at Bindukhera, when they are in full bloom.
If you can picture white tents under a canopy of orange-
coloured bloom; a multitude of brilliantly plumaged red and
gold minivets, golden orioles, rose-headed parakeets, golden-
backed woodpeckers, and wire-crested drongos flitting from
tree to tree and shaking down the bloom until the ground
round the tents resembled a rich orange-coloured carpet;
densely wooded foothills in the background topped by ridge
upon rising ridge of the Himalayas, and they in turn topped
by the eternal snows, then, and only then, will you have some
idea of our camp at Bindukhera one February morning in the
year 1929.

Bindukhera, which is only a name for the camping ground,
is on the western edge of a wide expanse of grassland some
twelve miles long and ten miles wide. When Sir Henry
Ramsay was king of Kumaon the plain was under intensive
cultivation, but at the time of my story there were only three
small villages, each with a few acres of cultivation dotted

117

along the banks of the sluggish stream that meanders down the length of the plain. The grass on the plain had been burnt a few weeks before our arrival, leaving islands of varying sizes where the ground was damp and the grass too green to burn. It was on these islands that we hoped to find the game that had brought us to Bindukhera for a week's shooting. I had shot over this ground for ten years and knew every foot of it, so the running of the shoot was left to me.

Shooting from the back of a well-trained elephant on the grasslands of the Tarai is one of the most pleasant forms of sport I know of. No matter how long the day may be, every moment of it is packed with excitement and interest, for in addition to the variety of game to be shot—on a good day I have seen eighteen varieties brought to bag ranging from quail and snipe to leopard and swamp deer—there is a great wealth of bird life not ordinarily seen when walking through grass on foot.

There were nine guns and five spectators in camp on the first day of our shoot that February morning, and after an early breakfast we mounted our elephants and formed a line, with a pad elephant between each two guns. Taking my position in the centre of the line, with four guns and four pad elephants on either side of me, we set off due south with the flanking gun on the right—fifty yards in advance of the line— to cut off birds that rose out of range of the other guns and were making for the forest on the right. If you are ever given choice of position in a line of elephants on a mixed-game shoot select a flank, but only if you are good with both gun and rifle. Game put up by a line of elephants invariably try to break out at a flank, and one of the most difficult objects to hit is a bird or an animal that has been missed by others.

When the air is crisp and laden with all the sweet scents that are to be smelt in an Indian jungle in the early morning, it goes to the head like champagne, and has the same effect on birds, with the result that both guns *and* birds tend to be too quick off the mark A too eager gun and a wild bird do not produce a heavy bag, and the first few minutes of all glorious days are usually as unproductive as the last few minutes when muscles are tired and eyes strained. Birds were plentiful that morning, and, after the guns had settled down, shooting improved and in our first beat along the edge of  the forest we picked up five peafowl, three red jungle fowl, ten black partridge, four grey partridge, two bush quail, and three hare. A good *sambhar* had been put up but he gained the shelter of the forest before rifles could be got to bear on him.

Where a tongue of forest extended out on to the plain for a few hundred yards, I halted the line. This forest was famous for the number of peafowl and jungle fowl that were always to be found in it, but as the ground was cut up by a number of deep nullahs that made it difficult to maintain a straight line, I decided not to take the elephants through it, for one of the guns was inexperienced and was shooting from the back of an elephant that morning for the first time. It was in this forest—when Wyndham and I some years previously were looking for a tiger—that I saw for the first time a cardinal bat. These beautiful bats, which look like gorgeous butterflies as they flit from cover to cover, are, as far as I know, only to be found in heavy elephant-grass.

After halting the line I made the elephants turn their heads to the east and move off in single file. When the last elephant had cleared the ground over which we had just beaten, I again halted them and made them turn their heads to the north. We were now facing the Himalayas, and hanging in the sky directly in front of us was a brilliantly lit white cloud that looked solid enough for angels to dance on.

The length of a line of seventeen elephants depends on the ground that is being beaten. Where the grass was heavy

I shortened the line to a hundred yards, and where it was light I extended it to twice that length. We had beaten up to the north for a mile or so, collecting thirty more birds and a leopard, when a ground owl got up in front of the line. Several guns were raised and lowered when it was realized what the bird was. These ground owls, which live in abandoned pangolin and porcupine burrows, are about twice the size of a partridge, look white on the wing, and have longer legs than the ordinary run of owls. When flushed by a line of elephants they fly low for fifty to a hundred yards before alighting. This I believe they do to allow the line to clear their burrows, for when flushed a second time they invariably fly over the line and back to the spot from where they originally rose. The owl we flushed that morning, however, did not behave as these birds usually do, for after flying fifty to sixty yards in a straight line it suddenly started to gain height by going round and round in short circles. The reason for this was apparent a moment later when a peregrine falcon, flying at great speed, came out of the forest on the left. Unable to regain the shelter of its burrow the owl was now making a desperate effort to keep above the falcon. With rapid wing beats he was spiralling upwards, while the falcon on widespread wings was circling up and up to get above his quarry. All eyes, including those of the *mahouts*, were now on the exciting flight, so I halted the line.

It is difficult to judge heights when there is nothing to make a comparison with. At a rough guess the two birds had reached a height of a thousand feet, when the owl—still moving in circles—started to edge away towards the big white cloud, and one could imagine the angels suspending their dance and urging it to make one last effort to reach the shelter of their cloud. The falcon was not slow to see the object of this manoeuvre, and he too was now beating the air with his wings and spiralling up in ever-shortening circles. Would the owl make it or would he now, as the falcon approached nearer to him, lose his nerve and plummet down in a vain effort to reach mother earth and the sanctuary of his burrow? Field glasses were now out for those who needed them, and up and down the line excited exclamations—in two languages—were running.

'Oh! he can't make it.'

'Yes he can, he can.'

'Only a little way to go now.'

'But look, look, the falcon is gaining on him.' And then, suddenly, only one bird was to be seen against the cloud. Well done! well done! *Shahbash! shahbash!* The owl had made it, and while hats were being waved and hands were being clapped, the falcon in a long graceful glide came back to the *semul* tree from which he had started.

The reactions of human beings to any particular event are unpredictable. Fifty-four birds and four animals had been shot that morning—and many more missed—without a qualm or the batting of an eyelid. And now, guns, spectators, and *mahouts* were unreservedly rejoicing that a ground owl had escaped the talons of a peregrine falcon.

At the northern end of the plain I again turned the line of elephants south, and beat down along the right bank of the stream that provided irrigation water for the three villages. Here on the damp ground the grass was unburnt and heavy, and rifles were got ready, for there were many hog deer and swamp deer in this area, and there was also a possibility of putting up another leopard.

We had gone along the bank of the stream for about a mile, picking up five more peafowl, four cock florican—hens were barred—three snipe, and a hog deer with very good horns when the accidental (please turn your eyes away, Recording Angel) discharge of a heavy high-velocity rifle in the hands of a spectator sitting behind me in my howdah, scorched the inner lining of my left ear and burst the eardrum. For me the rest of that February day was torture. After a sleepless night I excused myself on the plea that I had urgent work to attend to (again, please, Recording Angel) and at dawn, while the camp was asleep, I set out on a twenty-five-mile walk to my home at Kaladhungi.

The doctor at Kaladhungi, a keen young man who had recently completed his medical training, confirmed my fears that my eardrum had been destroyed. A month later we moved up to our summer home at Naini Tal, and at the Ramsay Hospital I received further confirmation of this diagnosis from Colonel Barber, Civil Surgeon of Naini Tal. Days passed, and it became apparent that abscesses were forming in my head. My condition was distressing my two sisters as much as it was distressing me, and as the hospital was unable to do anything to relieve me I decided—much against the wishes of my sisters and the advice of Colonel Barber—to go away.

I have mentioned this 'accident' not with the object of enlisting sympathy but because it has a very important bearing on the story of the Talla Des man-eater which I shall now relate.

2

Bill Baynes and Ham Vivian were Deputy Commissioners of, respectively, Almora and Naini Tal in the year 1929, and both were suffering from man-eaters, the former from the Talla Des man-eating tiger, and the latter from the Chowgarh man-eating tiger.

I had promised Vivian that I would try to shoot his tiger first, but as it had been less active during the winter months than Baynes's, I decided, with Vivian's approval, to try for the other first. The pursuit of this tiger would, I hoped, tide me over my bad time and enable me to adjust myself to my new condition. So to Talla Des I went.

My story concerns the Talla Des tiger, and I have refrained from telling it until I had written *Jungle Lore*. For without first reading *Jungle Lore*, and knowing that I had learnt— when a boy and later—how to walk in a jungle and use a rifle, the credulity of all who were not present in Kumaon at that time would have been strained and this, after my previous stories had been accepted at their face value, was the last thing I desired.

My preparations were soon made and on 4 April I left Naini Tal accompanied by six Garhwalis, among whom were Madho Singh and Ram Singh, a cook named Elahai, and a Brahmin, Ganga Ram, who did odd jobs and was very keen to go with me. Walking the fourteen miles down to Kathgodam we caught the evening train and, travelling through Bareilly and Pilibhit, arrived at noon next day at Tanakpur. Here I was met by the *peshkar*, who informed me that a boy had been killed the previous day by the Talla Des man-eater, and that under Baynes's orders two young buffalo—to be used as bait—had been dispatched for me via Champawat to Talla Des. After my men had cooked and eaten their food and I had breakfasted at the dak bungalow, we started off in good heart to try to walk the twenty-four miles to Kaladhunga (not to be confused with Kaladhungi) the same night.

The first twelve miles of the road—through Baramdeo to the foot of the sacred Purnagiri mountain—runs through forest most of the way. At the foot of the mountain the road ends, and there is the choice of two tracks to Kaladhunga. One, the longer, goes steeply up the left-hand side of the mountain to the Purnagiri temples, over a shoulder of the mountain, and down to Kaladhunga. The other track follows the alignment of the tramway line made by Collier when extracting the million cubic feet of *sal* timber that I have already spoken of. Collier's tramline—where it ran for four miles through the Sarda river gorge—has long since been washed away, but portions of the track he blasted across the perpendicular rock face of the mountain still remain. The going over this portion of the track was very difficult for my heavily-laden Garhwalis, and night came on when we were only halfway through the gorge. Finding a suitable place on which to camp for the night was not easy, but after rejecting

I

several places made dangerous by falling stones we eventually found a narrow shelf where the overhanging rock offered a measure of safety. Here we decided to spend the night, and after I had eaten my dinner and while the men were cooking their food with driftwood brought up from the river I undressed and lay down on my camp bed, the only article of camp equipment, excluding a washbasin and a forty-pound tent, that I had brought with me.

The day had been hot and we had covered some sixteen miles since detraining at Tanakpur. I was comfortably tired and was enjoying an after-dinner cigarette, when on the hill on the far side of the river I suddenly saw three lights appear. The forests in Nepal are burnt annually, the burning starting in April. Now, on seeing the lights, I concluded that the wind blowing down the gorge had fanned to flame the smouldering embers in some dead wood. As I idly watched these fires two more appeared a little above them. Presently the left-hand one of these two new fires moved slowly down the hill and merged into the central one of the original three. I now realized that what I had assumed were fires, were not fires but lights, all of a uniform size of about two feet in diameter, burning steadily without a flicker or trace of smoke. When presently more lights appeared, some to the left and others farther up the hill, an explanation to account for them presented itself. A potentate out on *shikar* had evidently lost some article he valued and had sent men armed with lanterns to search for it. Admittedly a strange explanation, but many strange things happen on the far side of that snow-fed river.

My men were as interested in the lights as I was, and as the river below us flowed without a ripple and the night was still, I asked them if they could hear voices or any other sounds—the distance across was about a hundred and fifty yards— but they said they could hear nothing. Speculation as to what was happening on the opposite hill was profitless, and as we were tired out after our strenuous day the camp was soon wrapped in slumber. Once during the night a *ghooral* sneezed in alarm on the cliff above us, and a little later a leopard called.

A long march and a difficult climb lay before us. I had warned my men that we would make an early start, and light was just showing in the east when I was given a cup of hot tea. Breaking camp, when only a few pots and pans had to be put away and a camp bed dismantled, was soon accomplished. As the cook and my Garhwalis streamed off in single file down a goat track into a deep ravine, which in Collier's day had been spanned by an iron bridge, I turned my eyes to the hill on which we had seen the lights. The sun was not far from rising and distant objects were now clearly visible. From crest to water's edge and from water's edge to crest I scanned every foot of the hill, first with my naked eyes and then with field glasses. Not a sign of any human being could I see, nor, reverting to my first theory, was there any smouldering wood, and it only needed a glance to see that the vegetation in this area had not been burnt for a year. The hill was rock from top to bottom, a few stunted trees and bushes growing where roothold had been found in crack or cranny. Where the lights had appeared was perpendicular rock where no human being, unless suspended from above, could possibly have gone.

Nine days later, my mission to the hill people accomplished, I camped for a night at Kaladhunga. For a lover of nature, or for a keen fisherman, there are few places in Kumaon to compare with Kaladhunga. From the bungalow Collier built when extracting the timber Nepal gave India, the land slopes gently down in a series of benches to the Sarda river. On these benches, where crops grew in the bygone days, there is now a luxuriant growth of grass. Here *sambhar* and cheetal are to be seen feeding morning and evening, and in the beautiful forests behind the bungalow live leopards and tigers, and a wealth of bird life including peafowl, jungle fowl, and *kalege* pheasants. In the big pools and runs below the bungalow some of the best fishing in the

Sarda river is to be had, either on a spinning rod with plug
bait or on a light rod with salmon fly or fly spoon.

At crack of dawn next morning we left Kaladhunga,
Ganga Ram taking the mountain track to Purnagiri and the
rest of us the shorter way through the Sarda gorge. Ganga
Ram's mission—which would entail an additional ten miles'
walk—was to present our thank-offerings to the sacred Purnagiri shrine. Before he left me I instructed him to find out all
he could, from the priests who served the shrine, about the
lights we had seen when on our way up to Talla Des. When
he rejoined me that evening at Tanakpur he gave me the
following information, which he had gleaned from the priests
and from his own observations.

Purnagiri, dedicated to the worship of the Goddess Bhag-
batti and visited each year by tens of thousands of pilgrims, is
accessible by two tracks. These, one from Baramdeo and the
other from Kaladhunga, meet on the northern face of the
mountain a short distance below the crest. At the junction
of the tracks is situated the less sacred of the two Purnagiri
shrines. The more sacred shrine is higher up and to the left.
This holy of holies can only be reached by going along a
narrow crack, or fault, running across the face of a more or
less perpendicular rock cliff. Nervous people, children, and
the aged are carried across the cliff in a basket slung on the
back of a hillman. Only those whom the Goddess favours are
able to reach the upper shrine; the others are struck blind and
have to make their offerings at the lower shrine.

Puja (prayer) at the upper shrine starts at sunrise and ends
at midday. After this hour no one is permitted to pass the
lower shrine. Near the upper and more sacred shrine is a
pinnacle of rock a hundred feet high, the climbing of which
is forbidden by the Goddess. In the days of long ago a
sadhu, more ambitious than his fellows, climbed the pinnacle
with the object of putting himself on an equality with the
Goddess. Incensed at his disregard of her orders, the Goddess hurled the *sadhu* from the pinnacle to the hill on the far

side of the snow-fed river. It is this *sadhu* who, banished for
ever from Purnagiri, worships the Goddess two thousand
feet above him by lighting lamps to her. These votive lights
only appear at certain times (we saw them on 5 April) and
are only visible to favoured people. This favour was accorded
to me and to the men with me, because I was on a mission to
the hillfolk over whom the Goddess watches.

That in brief was the information regarding the lights
which Ganga Ram brought back from Purnagiri, and im-
parted to me while we were waiting for our train at Tanakpur. Some weeks later I received a visit from the Rawal
(High Priest) of Purnagiri. He had come to see me about an
article I had published in a local paper on the subject of the
Purnagiri lights, and to congratulate me on being the only
European ever to have been privileged to see them. In my
article I gave the explanation for the lights as I have given it
in these pages, and I added that if my readers were unable to
accept this explanation and desired to find one for themselves, they should bear the following points in mind:

The lights did not appear simultaneously.
They were of a uniform size (about two feet in diameter).
They were not affected by wind.
They were able to move from one spot to another.

The High Priest was emphatic that the lights were an
established fact which no one could dispute—in this I was
in agreement with him for I had seen them for myself—and
that no other explanation than the one I had given could be
advanced to account for them.

The following year I was fishing the Sarda with Sir
Malcolm (now Lord) Hailey, who was Governor of the
United Provinces at the time. Sir Malcolm had seen my
article and as we approached the gorge he asked me to point
out the spot where I had seen the lights. We had four
dhimas (fishermen) with us who were piloting the *sarnis*
(inflated skins) on which we were floating down the river

from one fishing stand to the next. These men were part of a gang of twenty engaged by a contractor in floating pine sleepers from the high-level forests in Kumaon and Nepal to the boom at Baramdeo. This was a long, difficult, and very dangerous task, calling for great courage and a thorough knowledge of the river and its many hazards.

Below the shelf blasted out of the cliff by Collier, on which my men and I had spent the night when on our way up to Talla Des, was a narrow sandy beach. Here the *dhimas* at my request brought the *sarnis* to the bank, and we went ashore. After I had pointed out where the lights had appeared, and traced their movements on the hill, Sir Malcolm said the *dhimas* could possibly provide an explanation, or at least throw some light on the subject. So he turned to them—he knew the correct approach to make to an Indian when seeking information and could speak the language perfectly—and elicited the following information. Their homes were in the Kangra Valley where they had some cultivation, but not sufficient to support them. They earned their living by floating sleepers down the Sarda river for Thakur Dan Singh Bist. They knew every foot of the river as far down as Baramdeo, for they had been up and down it countless times. They knew this particular gorge very well, for there were backwaters in it that hung up the sleepers and gave them a great deal of trouble. They had never seen anything unusual in this part of the river in the way of lights, or anything else.

As he turned away from the *dhimas* I asked Sir Malcolm to put one more question to them. Had they in all the years they had been working on the Sarda ever spent a night in the gorge? Their answer to this question was a very emphatic No! Questioned further they said that not only had they never spent a night in the gorge but that they had never heard of anyone else ever having done so. The reason they gave for this was that the gorge was haunted by evil spirits.

Two thousand feet above us a narrow crack, worn smooth

by the naked feet of generations upon generations of devotees, ran for fifty yards across a perpendicular rock cliff where there was no handhold of any kind. In spite of the precautions taken by the priests to safeguard the lives of pilgrims, casualties while negotiating that crack were heavy until H.H. The Maharaja of Mysore provided funds a few years ago for a steel cable to be stretched across the face of the cliff, from the lower shrine to the upper.

So there well might be spirits at the foot of that cliff but not, I think, evil ones.

3

Now to get back to my story.

Ganga Ram, who could cover the ground as fast as any man in Kumaon, had stayed back with me to carry my camera, and we caught up with the cook and the six Garhwalis two miles from where we had spent the night. For the next six hours we walked with never a pause, at times through dense forests and at times along the bank of the Sarda river. Our way took us through Kaladhunga and through Chuka to the foot of the mountain, on the far side of which was our objective, the hunting grounds of the Talla Des man-eater. At the foot of the mountain we halted for two hours—to cook and eat our midday meal—before essaying the four-thousand-foot climb.

In the afternoon, with the hot April sun blazing down on our backs and without a single tree to shade us, we started on one of the steepest and most exhausting climbs my men and I had ever undertaken. The so-called road was only a rough track which went straight up the face of the mountain without a single hairpin bend to ease the gradient. After repeated and many halts we arrived at sunset at a little hamlet, a thousand feet from the crest. We had been warned at Chuka to avoid this hamlet, for, being the only inhabited place on the southern face of the mountain, it was visited regularly by the man-eater. However, man-eater or no man-eater, we could go no farther, so to the hamlet—which was a few hundred yards from the track—we went. The two families in the hamlet were delighted to see us, and after we had rested and eaten our evening meal, my men were provided with accommodation behind locked doors, while I settled down on my camp bed under a tree that sheltered the tiny spring which provided the two families with drinking water, with a rifle and a lantern to keep me company.

Lying on my bed that night I had ample time to review the situation. Instructions had been issued by Bill Baynes to headmen of villages not to disturb any human or other kills, pending my arrival. The boy the *peshkar* of Tanakpur had told me about, had been killed on the fourth and it was now the night of the sixth. Since leaving the train at Tanakpur we had not spared ourselves in an effort to try to get to the scene of the killing with as little delay as possible. I knew the tiger would have eaten out his kill before our arrival and that, if he was not disturbed, he would probably remain in the vicinity for a day or two. I had hoped when leaving camp that morning that we would reach our destination in time to tie out one of the young buffaloes, but the climb up from the Sarda had been too much for us. Regrettable as the loss of one day was, it could not be helped, and I could only hope that, if the tiger had moved away from the scene of his kill, he had not gone far. One of the disadvantages I had to contend with was that I did not know this part of Kumaon. The tiger had been operating for eight years and had made one hundred and fifty human kills, so it was reasonable to assume he was working over a very large area. If contact with him was once lost it might be weeks before it could again be made. However, worrying over what the tiger had done, or what he might do, was profitless, so I went to sleep.

I was to make an early start and it was still quite dark when Ganga Ram roused me by lighting the lantern which had gone out during the night. While breakfast was being got ready I had a bath at the spring, and the sun was just rising over the Nepal mountains when, having cleaned and oiled my .275 Rigby Mauser rifle and put five rounds in the magazine, I was ready to start. Inter-village communication had been interrupted by the man-eater and the two men in the hamlet had not heard about the tiger's last kill, so they were unable to give me any information as to the direction, or the distance, we would have to go. Not knowing when my men would get their next meal I told them to have a good one now and to follow me when they were ready, keeping close together and selecting open places to sit down in when they wanted to rest.

Rejoining the track up which we had laboured the previous evening, I halted for a spell to admire the view. Below me the valley of the Sarda was veiled in shadow and a wisp of mist showed where the river wound in and out through the foot-hills to emerge at Tanakpur. Beyond Tanakpur the eye could follow the river as a gleaming silver ribbon, until lost to sight on the horizon. Chuka was in shadow and partly obscured by mist, but I could see the path winding up to Thak, every foot of which I was to know when hunting the Thak man-eater ten years later. Thak village, gifted hundreds of years ago by the Chand Rajas of Kumaon to the priests who serve the Purnagiri shrines, was bathed in the morning sun, as was also the pinnacle of Purnagiri.

Twenty-five years have come and gone since I turned away from that view to complete the last stage of my journey to Talla Des—a long period, in which much has happened. But time does not efface events graven deep on memory's tablets, and the events of the five days I spent hunting the man-eating tiger of Talla Des are as clear-cut and fresh in my memory today as they were twenty-five years ago.

On the far side of the hill I found the track that I was on joined a quite good forest road some six feet wide, running east and west. Here I was faced with a dilemma, for there were no villages in sight and I did not know in which direction to go. Eventually, on the assumption that the road to the east could only take me out of my way as far as the Sarda, I decided to try it first.

Given the option of selecting my own time and place for a walk anywhere, I would unhesitatingly select a morning in early April on the northern face of a well-wooded hill in the Himalayas. In April all Nature is at her best; deciduous trees are putting out new leaves, each of a different shade of green or bronze; early violets, buttercups, and rhododendrons are giving way to later primulas, larkspurs, and orchids; and the birds—thrushes, babblers, minivets, tits, and a host of others —that migrated to the foothills for the winter are back on their nesting grounds and vie with each other in their joyous mating songs. Walking carefree and at ease in a forest in which there is no danger, only those objects and sounds which please the senses are looked at and listened to with any degree of attention, and all the other less-arresting sights and sounds blend together to form a pleasing whole. When there is danger from a man-eating tiger, however, the carefree feeling gives way to intense awareness.

Danger not only adds zest to all forms of sport, it also tends to sharpen the faculties and to bring into focus all that is to be seen and heard in a forest. Danger that is understood, and which you are prepared to face, does not detract in any way from pleasure. The bank of violets does not lose any of its beauty because the rock beyond it may shelter a hungry tiger, and the song of the black-headed sibia, poured out from the topmost branch of an oak tree, is none the less pleasing because a scimetar-babbler at the foot of the tree is warning the jungle folk of the presence of danger.

Fear may not be a heritage to some fortunate few, but I am not of their number. After a lifelong acquaintance with wild life I am no less afraid of a tiger's teeth and claws today than I was the day that a tiger shooed Magog and me out of the jungle in which he wanted to sleep. But to counter that fear and hold it in check I now have experience that I lacked in those early years. Where formerly I looked for danger all round me and was afraid of every sound I heard, I now knew where to look for danger, and what sounds to ignore or pay special attention to. And, further, where there was uncertainty where a bullet would go, there was now a measure of certainty that it would go in the direction I wanted it to. Experience engenders confidence, and without these two very important assets the hunting of a man-eating tiger on foot, and alone, would be a very unpleasant way of committing suicide.

The forest road I was walking on that April morning ran through an area in which a man-eating tiger was operating and had been used by the tiger frequently, as was evident from the scratch marks on it. In addition to these marks, none of which was fresh enough to show the pug-marks of the tiger which had made them, there were many tracks of leopard, *sambhar*, bear, *kakar*, and pig. Of birds there were many varieties, and of flowers there was great profusion, the most beautiful of which was the white butterfly orchid. These orchids hang down in showers and veil the branch or the trunk of the tree to which their roots are attached. One of the most artistic nests I have ever seen was that of a Himalayan black bear, made in a tree on which orchids were growing. A big oak tree had snapped off, either by weight of snow or in a storm, some forty feet above ground. Where the break had taken place a ring of branches, the thickness of a man's arm, had sprouted out at right angles to the trunk. Here moss had grown and in the moss butterfly orchids had found root-hold. It was here among these orchids that a bear had made its nest by bending over and pressing down the branches on to the broken-off tree trunk. The trees selected by bears in which to make their nests are of the variety whose branches will bend without snapping. The nests have nothing to do with family affairs and I have seen them at altitudes of from two thousand to eight thousand feet. At the lower altitude, to which bears descend during the winter months to feed on wild plums and honey, the nests give protection from ants and flies, and at the higher altitude they enable the animals to bask undisturbed in the sun.

When a road is interesting its length does not register on one's consciousness. I had been walking for about an hour when the forest ended and I came out on a grassy ridge over-looking a village. My approach over the open ground was observed, and when I reached the village the whole popula-tion appeared to have turned out to greet me. I often wonder whether in any other part of the world a stranger whose business was not known, arriving unexpectedly at a remote village, would be assured of the same welcome and hospitality as he would receive at any village throughout the length and breadth of Kumaon. I was possibly the first white man who had ever approached that village alone and on foot, and yet, by the time I reached the assembled people, a square of carpet had been produced, a *morha* (rush seat) placed on it, and I had hardly sat down before a brass vessel containing milk was placed in my hands. A lifelong associa-tion with the hillfolk enables me to understand the different dialects that are spoken in Kumaon and, what is just as important, to follow their every thought. As I had arrived armed with a rifle it was taken for granted that I had come to rid them of the man-eater, but what was puzzling them was my arrival on foot at that early hour when the nearest bungalow at which I could have spent the night was thirty miles away.

Cigarettes, passed round while I was drinking the milk, loosened tongues, and after I had answered the many questions put to me I put a few of my own. The name of the village, I learnt, was Tamali. The village had suffered for many years from the man-eater. Some said eight years and others said ten, but all were agreed that the man-eater had made its appearance the year that Bachi Singh had cut off his toes while splitting wood with an axe, and Dan Singh's black bullock, for which he had paid thirty rupees, had fallen down the hill and got killed. The last person killed at

Tamali by the man-eater had been Kundan's mother. She had been killed on the twentieth day of the previous month (March), while working with other women in a field below the village. No one knew whether the tiger was a male or a female, but all knew it was a very big animal, the fear of which was now so great that the outlying fields were no longer being cultivated and no one was willing to go to Tanakpur to get the food that was needed for the village. The tiger was never absent from Tamali for long, and if I stayed with them, which they begged me to do, I would have a better chance of shooting it than anywhere else in Talla Des.

To leave people who place implicit trust in you to the tender mercies of a man-eater is not easy. However, my reason for doing so was accepted, and, after I had assured the fifty or more people gathered round me that I would return to Tamali at the first opportunity, I bade them goodbye and set off to try to find the village where the last kill had taken place.

At the point where the track from the hamlet met the forest road I removed the sign I had placed on the road to indicate to my men that I had gone east, and replaced it on the road to the west, and, to ensure that there would be no mistake, I put a 'road closed' sign on the road to the east. The two signs I have mentioned are known throughout the hills, and, though I had not told my men that I would use them, I knew they would understand that I had laid them and would interpret them correctly. The first sign consists of a small branch laid in the middle of the road, held in position with a stone or bit of wood, with the leaves pointing in the direction in which it is intended that the person following should go.

The second sign consists of two branches crossed, in the form of an X.

The road to the west was level most of the way and ran through a forest of giant oak trees standing knee-deep in bracken and maidenhair fern. Where there were openings in this forest there were magnificent views of hills upon rising hills backed by the snowy range extending to east and west as far as the eye could see.

4

After going for some four miles due west the forest road turned to the north and crossed the head of a valley. Flowing down the valley was a crystal-clear stream which had its birth in the dense oak forest on the hill that towered above me on my left. Crossing the stream on stepping-stones, and going up a short rise, I came out on an open stretch of ground on the far side of which was a village. Some girls coming down from the village on their way to the stream caught sight of me as I came out on the open ground, and they called out in great excitement, 'The Sahib has come! The Sahib has come!' The cry was caught up from house to house and before I reached the village I was surrounded by an excited throng of men, women, and children.

From the headman I learnt that the name of the village was Talla Kote. That a *patwari* had arrived two days previously (5 April) from Champawat, to meet me and to tell all the people in the district that a sahib was coming from Naini Tal to try to shoot the man-eater. That shortly after the arrival of the *patwari* a woman of the village had been killed by the man-eater, and that in obedience to orders received from the Deputy Commissioner, Almora, the kill had not been disturbed. And finally, that in anticipation of my arrival a party of men had been sent that morning to look for the kill and, if there was anything of it left, to put up a *machan* for me. While the headman was giving me this information the party, numbering some thirty men, returned.

These men told me that they had searched the ground where the tiger had eaten its kill and that all they had been able to find were the woman's teeth. Even her clothes, they said, were missing. When I asked where the kill had taken place, a lad of about seventeen who was with the party of men said that if I would accompany him to the other side of the village he would point out to me where his mother had been killed by the man-eater. With the lad leading and the throng of men, women, and children following me, we went through the village to a narrow saddle some fifty yards long connecting two hills. This saddle was the apex of two great valleys. The one on the left, or western side, swept down towards the Ladhya river; the one on the right fell steeply away and down ten or fifteen miles to the Kali river. Halting on the saddle the lad turned and faced the valley on the right. The left-hand, or northern, side of this valley was under short grass with an odd bush scattered here and there, and the right-hand side was under dense scrub and tree jungle. Pointing to a bush on the grassy side eight hundred to a thousand yards away and a thousand to fifteen hundred feet below us, the lad said his mother had been killed near it while cutting grass in company with several other women. Then pointing to an oak tree in the ravine, the branches of which had been broken by *langurs*, he said it was under that tree that they had found the remains of his mother. Of the tiger, he said, neither he nor any of the party of men with him had seen or heard anything, but that when they were on their way down the hill they had heard first a *ghooral*, and then a little later, a *langur* calling.

A *ghooral* and a *langur* calling. *Ghooral* do occasionally call on seeing human beings, but not *langurs*. Both will call on seeing a tiger, however. Was it possible that the tiger had lingered near the scene of its kill and on being disturbed by the party of men had moved off and been seen, first by the *ghooral*, and then by the *langur*? While I was speculating on this point, and making a mental map of the whole country that stretched before me, the *patwarı*, who had been having his food when I arrived, joined me. Questioned about the two young buffaloes for which I had asked Baynes, the *patwari* said he had started out with them from Champawat and that he had left them at a village ten miles from Talla Kote, where a boy had been killed by the man-eater on 4 April within sight of the village. As there was no one on the spot to deal with the man-eater, the body had been recovered, and after a report of the occurrence had been sent to Champawat, from where it had been telegraphed to Tanak-pur for my information, he had given orders for the body of the boy to be cremated.

My men had not yet arrived from the hamlet where we spent the night, so, after instructing the headman to have my tent pitched on the open ground near the stream, I decided to go down and have a look at the ground where the tiger had eaten his kill, with the object of finding out if the man-eater was male or female, and if the latter, whether she had cubs. This part of Kumaon was, as I have already said, unknown to me, and when I asked the headman if he could tell me the easiest way to get down into the valley the lad, who had pointed out to me where his mother had been killed and eaten, stepped forward and said very eagerly, 'I will come with you, Sahib, and show you the way.'

The courage of people living in an area in which there is danger from a man-eater, and the trust they are willing to place in absolute strangers, has always been a marvel to me. The lad, whose name I learnt was Dungar Singh, was yet another example of that courage and trust. For years Dungar Singh had lived in fear of the man-eater and only an hour previously he had seen the pitiful remains of his mother. And yet, alone and unarmed, he was willing to accompany an absolute stranger into an area in which he had every reason to believe—from the alarm call of a *ghooral* and a *langur*—that the killer of his mother was lurking. True, he had only recently visited that area, but on that occasion he had been accompanied by thirty of his friends, and in numbers there was safety.

There was no way down the steep hillside from the saddle, so Dungar Singh led me back through the village to where there was a goat track. As we went down through scattered bushes I told him that my hearing was defective, that if he wanted to draw my attention to any particular thing to stop and point to it, and that if he wanted to communicate with me to come close and whisper into my right ear. We had gone about four hundred yards when Dungar Singh stopped and looked back. Turning round and looking in the same direction, I saw the *patwari* followed by a man carrying a shotgun hurrying down the hill after us. Thinking they had important information for me, I awaited their arrival and was disappointed to find that all the *patwari* wanted was to accompany me with his gun-bearer. This, very reluctantly, I permitted him to do for neither he nor his gun-bearer—both of whom were wearing heavy boots—looked like men who could move in a jungle without making considerable noise.

We had gone another four hundred yards through dense scrub jungle, when we came out on a clearing a few yards square. Here, where the goat track divided, one arm going towards a deep ravine on the left while the other followed the contour of the hill to the right, Dungar Singh stopped, and pointing in the direction of the ravine whispered that it was down there that the tiger had eaten his mother. As I did not wish the ground on which I wanted to look for pug-marks to be disturbed by booted men, I told Dungar Singh to stay on the open ground with the two men, while I went down alone into the ravine. As I stopped talking Dungar Singh whipped round and looked up the hill. When I looked in the same direction I saw a crowd of men standing on the saddle of the hill, where I had stood a little while before. With a hand stretched out towards us to ensure silence, and the other cupped to his ear, Dungar Singh was intently listening,

K

occasionally nodding his head. Then with a final nod he turned to me and whispered. 'My brother says to tell you that in the *wyran* field below you, there is something red lying in the sun.'

A *wyran* field is one that has gone out of cultivation, and below us on such a field there was something red lying in the sun. Maybe the red object was only a bit of dry bracken, or a *kakar* or young *sambhar*, but it might be a tiger. Anyway, I was not going to risk spoiling what might turn out to be a heaven-sent chance. So, handing my rifle to Dungar Singh, I took the *patwari* and his man, each by an arm, and led them to a medlar tree growing near by. Unloading the *patwari's* gun and laying it under a bush, I told the two men to climb the tree and on pain of death to remain quietly in it until I ordered them to come down. I do not think any two men ever climbed into a tree more gladly and from the way they clung to the branches after they had climbed as high as it was safe to go, it was evident that their views on man-eater hunting had undergone a drastic change since they followed me from the village.

The goat track to the right led on to a terraced field which had long been out of cultivation, and on which there was a luxuriant growth of oat grass. This field, about a hundred yards long, was ten feet wide at my end and thirty feet wide at the other, where it terminated on a ridge. For fifty yards the field was straight and then it curved to the left. As Dungar Singh saw me looking at it, he said that from the farther end we would be able to see down on to the *wyran* field on which his brother had seen the red object. Bending down and keeping to the inner edge of the field we crept along until we came to the far end. Here we lay down, and, crawling on hands and knees to the edge of the field, parted the grass and looked down.

Below us was a small valley with, on the far side, a steep grassy slope fringed on the side farthest from us by a dense growth of oak saplings. Beyond the saplings was the deep ravine in which the man-eater had eaten Dungar Singh's mother. The grassy slope was about thirty yards wide and below it was a rock cliff which, judging from the trees growing at the foot, was from eighty to a hundred feet high. On the near side of the slope was a terraced field, a hundred yards long and some ten yards wide. The field, which was in a straight line with us, had a small patch of short emerald-green grass at our end. On the remainder was a dense growth of an aromatic type of weed which grows to a height of four or five feet and has leaves like chrysanthemums, the undersides of which are white. Lying in brilliant sunlight on the patch of grass, and about ten feet apart, were two tigers.

The nearer tiger had its back to us with its head towards the hill, and the farther one had its stomach to us with its tail towards the hill. Both were fast asleep. The nearer offered the better shot, but I was afraid that on hearing the smack of the bullet the farther one would go straight down the hill into dense cover, in the direction in which its head was pointing. Whereas if I fired at the farther one first, the smack of the bullet—not to be confused with the crack of the rifle—would either drive the nearer one up the hill where there was less cover or else drive it towards me. So I decided to take the farther one first. The distance was approximately one hundred and twenty yards, and the angle of fire was not so steep that any allowance had to be made for the lift of the bullet, a point which has to be kept in mind when shooting downhill on the Himalayas. Resting the back of my hand on the edge of the field, to form a cushion, and holding the rifle steady, I took careful aim at where I thought the animal's heart would be and gently pressed the trigger. The tiger never moved a muscle, but the other one was up like a flash and in one bound landed on a five-foot-high bank of earth that divided the field from a rain-water channel. Here the second tiger stood, broadside on to me, looking back over its right shoulder at its companion. At my shot it reared up and fell over backwards into the rain-water channel, and out of sight.

After my second shot I saw a movement in the aromatic weeds which started close to where the dead tiger was lying. A big animal was going at full gallop straight along the field. Having started from so close to where the two tigers had been lying, this third animal could only be another tiger. I could not see the animal, but I could follow its movements by the parting of the weeds the leaves of which were white on the underside. Flicking up the two-hundred-yard leaf-sight I waited for the animal to break cover. Presently out on to the grassy slope dashed a tiger. I now noticed that the slope the tiger was on curved to the right, in the same way as the field I was lying on curved to the left. As the tiger was keeping to the contour of the hill this curve in the slope enabled me to get a near-broadside shot at it.

I have seen animals fall over at a shot, and I have seen them crumple up, but I have never seen an animal fall as convincingly dead as that tiger fell at my shot. For a few moments it lay motionless and then it started to slide down, feet foremost, gaining momentum as it went. Directly below it, and within a few feet of the brink of the rock cliff, was an oak sapling eight to ten inches thick. The tiger struck this sapling with its stomach and came to rest with its head and forelegs hanging down on one side and its tail and hindlegs hanging down on the other. With rifle to shoulder and finger on trigger I waited, but there was not so much as a quiver in the tiger. Getting to my feet I beckoned to the *patwari*, who from his seat on the medlar tree had obtained a grandstand view of the whole proceedings. Dungar Singh, who had lain near me breathing in short gasps, was now dancing with excitement and from the way he was glancing at the tigers and then up at the crowd of people on the saddle, I knew he was thinking of the tale he would have to tell that night and for many moons thereafter.

When I saw the two tigers lying asleep I concluded that

the man-eater had found a mate, but later, when my third shot flushed a third tiger, I knew I was dealing with a tigress and her two cubs. Which of the three was the mother and which the cubs it was not possible to say, for all three looked about the same size when I had viewed them over the sights of my rifle. That one of the three was the man-eater of Talla Des there could be no question, for tigers are scarce in the hills, and these three tigers had been shot close to where a human being had recently been killed and eaten. The cubs had died for the sins of their mother. They had undoubtedly eaten the human flesh their mother had provided for them from the time they were weaned; this, however, did not mean that when they left the protection of their mother they would have become man-eaters themselves. For in spite of all that has been said since *Man-eaters of Kumaon* was published, I still maintain that the cubs of man-eating tigers—in that part of India about which I am writing—do not become man-eaters simply because they have eaten human flesh when young.

Sitting on the edge of the field with my feet dangling down and the rifle resting on my knees, I handed cigarettes to my companions and told them I would go and have a look at the tiger that had fallen into the rain-water channel, after we had finished our smoke. That I would find the tiger dead I had no doubt whatever; even so, nothing would be lost by waiting a few minutes, if for no other reason than to give myself a little time to rejoice over the marvellous luck I had met with. Within an hour of my arrival at Talla Des I had, quite by accident, got in touch with a man-eater that had terrorized an area of many hundreds of square miles for eight years, and in a matter of a few seconds had shot dead the man-eater and her two cubs. To the intense pleasure that all sportsmen feel at having held a rifle steady when every drop of blood in one's body was pounding with excitement, was added the pleasure and relief of knowing that there would be no necessity to follow up a wounded animal, a contingency that has to be faced when hunting tigers on foot.

My men would not ascribe my good fortune to luck. To avoid the possibility of failure they had consulted the old priest at the temple in Naini Tal and he had selected the propitious day for us to start on our journey to Talla Des, and evil omens when we started had been absent. My success would not be ascribed to good luck, therefore; nor, if I had failed to shoot the tigers, would my failure have been ascribed to bad luck, for no matter how well aimed a bullet might be it could do no harm to an animal whose time to die had not come. The superstitions of those whom I have been associated with on *shikar* have always been of interest to me. Being myself unwilling to begin a journey on a Friday, I am not inclined to laugh at a hillman's rooted aversion to begin a journey to the north on Tuesday or Wednesday, to the south on Thursday, to the east on Monday or Saturday, or to the west on Sunday or Friday. To permit those who accompany one on a dangerous mission to select the day for the start of the journey is a small matter, but it makes all the difference between having cheerful and contented companions or companions who are oppressed by a feeling of impending disaster.

The four of us sitting on the edge of the field had nearly finished our cigarettes, when I noticed that the tiger that was resting against the oak sapling was beginning to move. The blood from the body had evidently drained into the forward end of the animal, making that end heavier than the tail end, and it was now slowly slipping down head foremost. Once it was clear of the sapling the tiger glissaded down the grassy slope, and over the brink of the rock cliff. As it fell through space I threw up the rifle and fired. I fired that shot on the spur of the moment to give expression to my joy at the success of my mission to Talla Des, and also, I am ashamed to admit, to demonstrate that there was nothing—not even a tiger falling through space—that I could not hit on a day like this. A moment after the tiger disappeared among the tree tops, there was a rending of branches, followed by a dull and heavy thud. Whether or not I had hit the falling tiger did not matter, but what did matter was that the men of the village would have farther to carry it now than if it had remained on the slope.

My cigarette finished, I told my companions to sit still while I went down to look at the tiger in the rain-water channel. The hill was very steep and I had climbed down some fifty feet when Dungar Singh called out in a very agitated voice. 'Look, Sahib, look. There goes the tiger.' With my thoughts on the tiger below me, I sat down and raised my rifle to meet the charge I thought was coming. On seeing my preparations, the lad called out, 'Not here, but there, Sahib, there.' Relieved of the necessity of guarding my front I turned my head and looked at Dungar Singh and saw he was pointing across the main valley to the lower slopes of the hill on which his mother had been killed. At first I could see nothing, and then I caught sight of a tiger going diagonally up towards a ridge that ran out from the main hill. The tiger was very lame and could only take three or four steps at a time, and on its right shoulder was a big patch of blood. The patch of blood showed it was the tiger that had crashed through the trees, for the tiger that had fallen into the rain-water channel had been shot in the left shoulder.

Growing on the hill close to where I was sitting was a slender pine sapling. Putting up the three-hundred-yard leaf-sight I got a firm grip of the sapling with my left hand and resting the rifle on my wrist took a careful and an unhurried shot. The distance was close on four hundred yards and the tiger was on a slightly higher elevation than I was, so, taking a very full sight, I waited until it again came to a stand and then gently pressed the trigger. The bullet appeared to take an incredibly long time to cover the distance, but at last I saw a little puff of dust and at the same moment the tiger lurched forward, and then carried on with its slow walk. I had taken a little too full a sight, and the bullet had gone a shade too high. I now had the range to a nicety and all that

I needed to kill the tiger was one more cartridge; the cartridge I had foolishly flung away when the tiger was falling through the air. With an empty rifle in my hands, I watched the tiger slowly and painfully climb to the ridge, hesitate for a few moments, and then disappear from view.

Sportsmen who have never shot in the Himalayas will question my wisdom in having armed myself with a light .275 rifle, and only carrying five rounds of ammunition. My reasons for having done so were:

(*a*) The rifle was one I had used for over twenty years, and with which I was familiar.

(*b*) It was light to carry, accurate, and sighted up to three hundred yards.

(*c*) I had been told by Colonel Barber to avoid using a heavy rifle, and not to fire more shots than were necessary with a light one.

With regard to ammunition, I had not set out that morning to shoot tigers but to find the village where the last human kill had taken place and, if I had the time, to tie out a young buffalo as bait. As it turned out, both the light rifle and the five rounds would have served my purpose if I had not thrown away that vital round.

My men arrived at the village in time to join the crowd on the saddle, and to witness the whole proceedings. They knew that the five rounds in the magazine of the rifle were all the ammunition I had with me, and when after my fifth shot they saw the wounded tiger disappear over the ridge, Madho

Singh came tearing down the hill with a fresh supply of ammunition.

The tiger on the patch of green grass, and the tiger in the rain-water channel—which I found lying dead where it had fallen—were both nearly full-grown, and the one that had got away wounded was quite evidently their mother, the man-eater of Talla Des. Leaving Madho Singh and Dungar Singh to make arrangements for the cubs to be carried up to the village, I set out alone to try to get in touch with the wounded tigress. From the bed of bracken on to which she had fallen after crashing through the trees, I followed a light blood-trail to where she had been standing when I fired my last shot. Here I found a few cut hairs clipped from her back by my bullet, and a little extra blood which had flowed from her wound when she lurched forward on hearing my bullet strike the ground above her. From this spot to the ridge there was only an occasional drop of blood, and on the short stiff grass beyond the ridge I lost the trail. Close by was a dense patch of scrub, a hundred yards wide, extending up the side of a steep hill for three hundred yards, and I suspected that the tigress had taken shelter in this scrub. But as night was now closing in and there was not sufficient light for accurate shooting, I decided to return to the village and leave the searching of the scrub until the following day.

5

The next morning was spent in skinning the cubs and in pegging out their skins with the six-inch nails I had brought with me from Naini Tal. While I was performing this task at least a hundred vultures alighted on the trees fringing the open ground on which my tent was pitched. It was these that brought to light the missing clothes of the man-eater's victim, for the cubs had torn the blood-soaked garments into strips and swallowed them.

The men of the village sat round me while I was skinning the cubs and I told them I wanted them to assist my

Garhwalis in beating out the patch of scrub in which I thought
the wounded tigress had taken shelter. This they were very
willing to do. At about midday we set off, the men going
through the village and along the saddle to the top of the hill
above the cover, while I went down the goat track into the
valley and up to the ridge over which I had followed the
tigress the previous evening. At the lower edge of the scrub
there was an enormous boulder the size of a small house.
Climbing to the top of the boulder—from which I was visible
to the men at the top of the hill—I waved my hat as a signal
for them to start the beat. To avoid the risk of anyone getting
mauled, I had instructed the men to stay on the top of the
hill and, after clapping their hands and shouting, to roll rocks
down the hillside into the scrub I have spoken of. One *kakar*
and a few *kalege* pheasants came out of the bushes, but
nothing else. When the rocks had searched out every foot of
the ground, I again waved my hat as a signal for the men to
stop the beat and return to the village.

When the men had gone I searched the cover, but without
any hope of finding the tigress. As I watched her going up
the hill the previous evening I could see that she was suffering
from a very painful wound, and when I examined the blood
where she had lurched forward, I knew the wound was a
surface one and not internal. Why then had the tigress fallen
to my bullet as if poleaxed, and why had she hung suspended
from the oak sapling for a matter of ten to fifteen minutes
without showing any signs of life? To these questions I
could not at the time nor can I now find any reasonable
answer. Later I found my soft-nose, nickel-encased bullet
firmly fixed in the ball-and-socket joint of the right shoulder.
When the flight of a high-velocity bullet is arrested by
impact with a bone the resulting shock to an animal is very
considerable. Even so, a tiger is a heavy animal with a
tremendous amount of vitality, and why a light .275 bullet
should knock such an animal flat and render it unconscious
for ten or fifteen minutes is to me inexplicable.

Returning to the ridge,
I stood and surveyed the
country. The ridge ap-
peared to be many miles
long and divided two val-
leys. The valley to the left
at the upper end of which
was the patch of scrub was
open grass country, while
the valley to the right at
the upper end of which the tigers had eaten the woman had
dense tree and scrub jungle on the right-hand side, and a
steep shaly slope ending in a rock cliff on the left.

Sitting down on a rock on the ridge to have a quiet smoke,
I reviewed the events of the previous evening, and came to
the following conclusions:

(a) From the time the tigress fell to my shot to the time
   she crashed through the trees, she had been unconscious.

(b) Her fall, cushioned by the trees and the bed of bracken,
   had restored consciousness but had left her dazed.

(c) In this dazed condition she had just followed her nose
   and on coming up against the hill she had climbed it
   without knowing where she was going.

The question that now faced me was: How far and in
what direction had the tigress gone? Walking downhill with
an injured leg is far more painful than walking uphill, and as
soon as the tigress recovered from her dazed condition she
would stop going downhill and would make for cover in
which to nurse her injury. To get to cover she would have to
cross the ridge, so the obvious thing was to try to find out if
she had done so. The task of finding if a soft-footed animal
had crossed a ridge many miles long would have been a
hopeless one if the ridge had not had a knife-edge. Running
along the top was a game track, with an ideal surface for

recording the passage of all the animals that used it. On the left of the track was a grassy slope and on the right a steep shale scree ending in a sheer drop into the ravine below.

Finishing my smoke I set off along the game track on which I found the tracks of *ghooral*, *sarao*, *sambhar*, *langur*, porcupine, and the pug-marks of a male leopard. The farther I went the more despondent I grew, for I knew that if I did not find the tigress's pug-marks on this track there was little hope of my ever seeing her again. I had gone about a mile along the ridge, disturbing two *ghooral* who bounded away down the grassy slope to the left, when I found the pug-marks of the tigress, and a spot of dry blood. Quite evidently, after disappearing from my view over the ridge the previous evening, the tigress had gone straight down the grassy slope until she recovered from her dazed condition and then had kept to the contour of the hill, which brought her to the game track. For half a mile I followed her pug-marks to where the shale scree narrowed to about fifteen yards. Here the tigress attempted to go down the scree, evidently with the intention of gaining the shelter of the jungle on the far side of the ravine. Whether her injured leg failed her or whether dizziness overcame her, I do not know; anyway, after falling forward and sliding head-foremost for a few yards she turned round and with legs widespread clawed the ground in a desperate but vain effort to avoid going over the sheer drop into the ravine below. I am as sure-footed as a goat, but that scree was far too difficult for me to attempt to negotiate, so I carried on along the track for a few hundred yards until I came to a rift in the hill. Down this rift I climbed into the ravine.

As I walked up the thirty-yard-wide ravine I noted that the rock cliff below the shale scree was from sixty to eighty feet high. No animal, I was convinced, could fall that distance on to rocks without being killed. On approaching the spot where the tigress had fallen I was overjoyed to see the white underside of a big animal. My joy, however, was

short lived, for I found the animal was a *sarao* and not the tigress. The *sarao* had evidently been lying asleep on a narrow ledge near the top of the cliff and, on being awakened by hearing, and possibly scenting, the tigress above him, had lost his nerve and jumped down, breaking his neck on the rocks at the foot of the cliff. Close to where the *sarao* had fallen there was a small patch of loose sand. On this the tigress had landed without doing herself any harm beyond tearing open the wound in her shoulder. Ignoring the dead *sarao*, within a yard of which she passed, the tigress crossed the ravine, leaving a well-defined blood trail. The bank on the right-hand side of the ravine was only a few feet high, and several times the tigress tried but failed to climb it. I knew now that I would find her in the first bit of cover she could reach. But my luck was out. For some time heavy clouds had been massing overhead, and before I found where the tigress had left the ravine a deluge of rain came on, washing out the blood trail. The evening was now well advanced and as I had a long and a difficult way to go, I turned and made for camp.

Luck plays an important part in all sport, and the tigress had—so far—had her full share of it. First, instead of lying out in the open with her cubs where I would have been able to recognize her for what she was, she was lying out of sight in thick cover. Then, the flight of my bullet had been arrested by striking the one bone that was capable of preventing it from inflicting a fatal wound. Later the tigress had twice fallen down a rock cliff, where she would undoubtedly have been killed had her fall in the one case not been cushioned by branches and a bed of bracken and in the other by a soft patch of sand. And finally, when I was only a hundred yards from where she was lying up, the rain came down and washed out the blood trail. However, I too had had a measure of luck, for my fear that the tigress would wander away down the grassy slope where I would lose touch with her had not been realized, and, further, I knew now where to look for her.

6

Next morning I returned to the ravine, accompanied by my six Garhwalis. Throughout Kumaon the flesh of *sarao* is considered a great delicacy, and as the young animal that had broken its neck was in prime condition, it would provide a very welcome meat ration for my men. Leaving the men to skin the *sarao*, I went to the spot from where I had turned back the previous evening. Here I found that two deep and narrow ravines ran up the face of the hill on the right. As it was possible that the tigress had gone up one of these, I tried the nearer one first only to find, after I had gone up it for a few hundred yards, that the sides were too steep for any tiger to climb, and that it ended in what in the monsoon rains must have been a thirty-foot-high waterfall. Returning to my starting point I called out to the men, who were about fifty yards away up the main ravine, to light a fire and boil a kettle of water for my tea. I then turned to examine the second ravine and as I did so I noticed a well-used game track coming down the hill on the left-hand side. On the game track I found the pug-marks of the tigress, partly obliterated by the rain of the previous evening. Close to where I was standing was a big rock. On approaching this rock I saw that there was a little depression on the far side. The dead leaves in the depression had been flattened down, and on them were big clots of blood. After her fall into the ravine—which may have been forty hours earlier—the tigress had come to this spot and had only moved off on hearing me call to the men to boil the kettle for tea.

Owing to differences in temperament it is not possible to predict what a wounded tiger will do when approached by a human being on foot, nor is it possible to fix a period during which a wounded tiger can be considered as being dangerous —that is, liable to charge when disturbed. I have seen a tiger with an inch-long cut in a hind pad, received while running away, charge full out from a distance of a hundred yards five minutes after receiving the wound; and I have seen a tiger that had been nursing a very painful jaw wound for many hours allow an approach to within a few feet without making any attempt to attack. Where a wounded man-eating tiger is concerned the situation is a little complicated, for, apart from not knowing whether the wounded animal will attack on being approached, there is the possibility—when the wound is not an internal one—of its attacking to provide itself with food. Tigers, except when wounded or when man-eaters, are on the whole very good-tempered. Were this not so it would not be possible for thousands of people to work as they do in tiger-infested jungles, nor would it have been possible for people like me to have wandered for years through the jungles on foot without coming to any harm. Occasionally a tiger will object to too close an approach to its cubs or to a kill that it is guarding. The objection invariably takes the form of growling, and if this does not prove effective it is followed by short rushes accompanied by terrifying roars. If these warnings are disregarded, the blame for any injury inflicted rests entirely with the intruder. The following experience with which I met some years ago is a good example of my assertion that tigers are good-tempered. My sister Maggie and I were fishing one evening on the Boar river three miles from our home at Kaladhungi. I had caught two small *mahseer* and was sitting on a rock smoking when Geoff Hopkins, who later became Conservator of Forests, Uttar Pradesh, turned up on his elephant. He was expecting friends, and being short of meat he had gone out with a .240 rook-rifle to try to shoot a *kakar* or a peafowl. I had caught all the fish we needed, so we fell in with Geoff's suggestion that we should accompany him and help him to find the game he was looking for. Mounting the elephant we crossed the river and I directed the *mahout* to a part of the jungle where *kakar* and peafowl were to be found. We were going through short grass and plum jungle when I caught sight of a dead *cheetal* lying under a tree. Stopping the elephant I slipped to the ground and went to see what had killed the *cheetal*. She was an old hind that had been dead for twenty-four hours, and as I could find no marks of injury on her I concluded that she had died of snake-bite. As I turned to rejoin the elephant I saw a drop of fresh blood on a leaf. The shape of the drop of blood showed that the animal from which it had come had been moving away from the dead *cheetal*. Looking a little farther in the direction in which the splash from the blood indicated the animal had gone, I saw another spot of blood. Puzzled by this fresh blood-trail I set off to see where it led to, and signalled to the elephant to follow me. After going over short grass for sixty or seventy yards the trail led towards a line of thick bushes some five feet high. Going up to the bushes where the trail ended I stretched out both arms—I had left my rod on the elephant—and parted the bushes wide, and there under my outstretched hands was a *cheetal* stag with horns in velvet, and lying facing me and eating the stag was a tiger. As I parted the bushes the tiger looked up and the expression on its face said, as clearly as any words, 'Well, I'll be damned!' Which was exactly what I was saying to myself. Fortunately I was so surprised that I remained perfectly still—possibly because my heart had stopped beating—and after looking straight into my face for a moment the tiger, who was close enough to have stretched out a paw and stroked my head, rose, turned, and sprang into the bushes behind him all in one smooth graceful movement. The tiger had killed the stag among the plum bushes shortly before our arrival, and in taking it to cover he went past the dead hind, leaving the blood trail that I followed. The three on the elephant did not see the tiger until he was in the air, when the *mahout* exclaimed with horror, 'Khabardar, Sahib. Sher hai.' He was telling me that it was a tiger and to be careful.

Rejoining my men I drank a cup of tea while they cut up the *sarao* into convenient bits to carry, and returned with them to the depression in which I had found the clots of

L

blood. All six men had been out on *shikar* with me on many occasions, and on seeing the quantity of blood they were of the opinion that the tigress had a body wound which would prove fatal in a matter of hours. On this point we were not in agreement, for I knew the wound was a superficial one from which the tigress, given time, would recover, and that the longer she lived the more difficult it would be to get in touch with her.

If you can imagine a deep and narrow ravine running up the face of a steep hill with the ground on the right sloping towards the ravine and well wooded but free of under-growth, and the ground on the left-hand side of the ravine sloping upwards and covered with dense patches of *ringal* (stunted bamboo), bracken, and brushwood of all kinds, you will have some idea of the country my men and I worked over for the rest of that day.

My plan was for the men to go up on the right-hand side of the ravine, to keep me in sight by climbing into the highest trees they could find, and, if they wished to attract my attention, to whistle—hillmen, like some boys, are very good at whistling through their teeth. They would be in no danger from the tigress, for there was no cover on their side, and all of them were expert tree-climbers. The tracks of the tigress after she left the depression near the big rock showed that she had gone up the hill on the left-hand side of the ravine. Up this hill I now started to follow her.

I have emphasized elsewhere that jungle lore is not a science that can be learnt from textbooks, but that it can be absorbed a little at a time, and that the absorption process can go on indefinitely. The same applies to tracking. Track-ing, because of its infinite variations, is one of the most interesting forms of sport I know, and it can at times be also the most exciting. There are two generally accepted methods of tracking. One, following a trail on which there is blood, and the other, following a trail on which there is no blood. In addition to these two methods I have also at times been able to find a wounded animal by following blowflies, or by following meat-eating birds. Of the two generally-accepted methods, following a blood-trail is the more sure way of finding a wounded animal. But as wounds do not always bleed, wounded animals have at times to be tracked by their foot-prints or by the disturbance to vegetation caused by their passage. Tracking can be easy or difficult according to the nature of the ground, and also according to whether the animal being tracked has hard hooves or soft pads. When the tigress left the depression—on hearing me calling to my men—her wound had stopped bleeding and the slight discharge that was coming from the wound owing to its having turned septic was not sufficient to enable me to follow her, so I had to resort to tracking her by her foot-prints and by disturbed vegetation. This, on the ground I was on, would not be difficult, but it would be slow, and time was on the side of the tigress. For the longer the trail the better the chance would be of her recovering from her wound and the less chance there would be of my finding her, for the strain of the past few days was now beginning to tell on me.

For the first hundred yards the trail led through knee-high bracken. Here tracking was easy, for the tigress had kept to a more or less straight line. Beyond the bracken was a dense thicket of *ringal*. I felt sure the tigress would be lying up in this thicket, but unless she charged there was little hope of my getting a shot at her, for it was not possible to move silently through the matted *ringals*. When I was halfway through the thicket a *kakar* started barking. The tigress was on the move, but instead of going straight up the hill she had gone out on the left, apparently on to open ground, for the *kakar* was standing still and barking. Retracing my steps I worked round to the left but found no open ground in that direction, nor did I appear to be getting any nearer the barking deer. The *kakar*, soon after, stopped barking and a number of *kalege* pheasants started chattering. The tigress was still on the move, but, turn my head as I would, I could not locate the sound.

Pin-pointing, that is, fixing the exact direction and distance of all sounds heard, is a jungle accomplishment which I have reduced to a fine art and of which I am very proud. Now, for the first time, I realized with a shock that my accident had deprived me of this accomplishment and that no longer would I be able to depend on my ears for safety and for the pleasure of listening intimately to the jungle folk whose language it had taken me years to learn. Had my remaining ear been sound it would not have mattered so much, but unfortunately the drum of that ear also had been injured by a gun 'accident' many years previously. Well, there was nothing that could be done about it now, and handicapped though I was I was not going to admit at this stage of the proceedings that any tiger, man-eater or other, had any advantage over me when we were competing for each other's lives under conditions that favoured neither side.

Returning to the bracken, I started to try to find the tigress, depending on my eyes only. The jungle appeared to be well stocked with game, and I repeatedly heard *sambhar*, *kakar*, and *langur* giving their alarm calls, and more than once I heard pheasants, jays, and white-capped laughing thrush mobbing the tigress. Paying no attention to these sounds, which ordinarily I would have listened for eagerly, I tracked the tigress foot by foot as, resting frequently, she made her way up the hill, at times in a straight line and at times zig-zagging from cover to cover. Near the top of the hill was a stretch of short stiff grass about a hundred yards wide. Beyond this open ground were two patches of dense brush-wood divided by a narrow lane which ran up to the top of the hill. On the short stiff grass I lost the tracks. The tigress knew she was being closely followed and would therefore expose herself as little as possible. The patch of brushwood to my right front was thirty yards nearer than the patch to the left, so I decided to try it first. When I was within a yard or two of the cover I heard a dry stick snap under the weight of some heavy animal. I was positive on this occasion that the sound had come from the left, so I turned and went to the patch of brushwood from which the sound appeared to have come. This was the second mistake I made that day—the first was calling to my men to boil the kettle for tea—for my men told me later that I crossed the open stretch of ground on the heels of the tigress, and that when I turned and walked away to the left she was lying on an open bit of ground a few yards inside the bushes, evidently waiting for me.

Finding no trace of the tigress in the brushwood on the left I came back to the open ground, and, on hearing my men whistling, looked in the direction in which I expected them to be. They had climbed to the top of a tree a few hundred yards to my right, and when I lifted my hand to indicate that I had seen them, they waved me up, up, up, and then down, down, down. They were letting me know that the tigress had climbed to the top of the hill, and that she had gone down on the far side. Making what speed I could I went up the narrow lane and on reaching the top found an open hillside. On this the grass had been burnt recently, and in the ashes, which were still damp from the rain of the previous evening, I found the pug-marks of the tigress. The hill sloped gently down to a stream, the one that I had crossed several miles higher up on the day of my arrival at Talla Kote. After lying down and quenching her thirst the tigress had crossed the stream and gone up into the thick jungle beyond. It was now getting late, so I retraced my steps to the top of the hill and beckoned to my men to join me.

From the big rock where I took up the tracks of the tigress to the stream where I left them was only some four miles, and it had taken me seven hours to cover the distance. Though it had ended in failure the day had been an interesting and exciting one. Not only for me who, while doing the tracking, had to avoid being ambushed by a wounded man-eating tiger, but also for my Garhwalis who by climbing trees had kept both the tigress and myself in view most of the time. And it had been a long day also, for we had started at daylight, and it was 8 p.m. when we got back to camp.

7

The following morning while my men were having their food I attended to the skins, re-pegging them on fresh ground and rubbing wood ashes and powdered alum on the damp parts. Tiger skins need a lot of care, for if every particle of fat is not removed and the lips, ears, and pads properly treated, the hair slips, ruining the skin. A little before midday I was ready to start, and accompanied by four of my men—I left the other two men in camp to attend to the *sarao's* skin—I set out for the place where I had stopped tracking the tigress the previous evening.

The valley through which the stream flowed was wide and comparatively flat, and ran from west to east. On the left-hand side of the valley was the hill on the far side of which I had followed the tigress the previous day, and on the right-hand side was the hill along which ran the road to Tanakpur. Before the advent of the man-eater the valley between these two hills had been extensively grazed over by the cattle of Talla Kote, and in consequence the ground was criss-crossed by a maze of cattle paths, and cut up with narrow eroded water-channels. Dotted about the valley were open glades of varying sizes surrounded by dense scrub and tree jungle. Good ground on which to hunt *sambhar*, *kakar*, and bear, all of whose tracks were to be seen on the cattle paths, but not the ground one would select on which to hunt a man-eating tiger. The hill on the left commanded an extensive view of the valley, so I spaced my men in trees along the crest at intervals of two hundred yards to keep a look-out and to be on hand in case they were needed. I then went down to the spot where I had left the tracks of the tigress the previous evening.

I had wounded the tigress on 7 April, and it was now the 10th. As a general rule a tiger is not considered to be dangerous—that is, liable to charge at sight—twenty-four hours after being wounded. A lot depends on the nature of the wound, however, and on the temper of the wounded individual. Twenty-four hours after receiving a light flesh wound a tiger usually moves away on being approached, whereas a tiger with a painful body-wound might continue to be dangerous for several days. I did not know the nature of the wound the tigress was suffering from, and as she had made no attempt to attack me the previous day I believed I could now ignore the fact that she was wounded and look upon her only as a man-eater, and a very hungry man-eater at that, for she had eaten nothing since killing the woman whom she had shared with the cubs.

Where the tigress had crossed the stream there was a channel, three feet wide and two feet deep, washed out by rain-water. Up this channel, which was bordered by dense brushwood, the tigress had gone. Following her tracks I came to a cattle path. Here she had left the channel and gone along the path to the right. Three hundred yards along was a tree with heavy foliage and under this tree the tigress had lain all night. Her wound had troubled her and she had tossed about, but on the leaves on which she had been lying there was neither blood nor any discharge from her wound. From this point on I followed her fresh tracks, taking every precaution not to walk into an ambush. By evening I had tracked her for several miles along cattle paths, water channels, and game tracks, without having set eyes on so much as the tip of her tail. At sunset I collected my men, and as we returned to camp they told me they had been able to follow the movements of the tigress through the jungle by the animals and birds that had called at her, but that they too had seen nothing of her.

When hunting unwounded man-eating tigers the greatest danger, when walking into the wind, is of an attack from behind, and to a lesser extent from either side. When the wind is from behind, the danger is from either side. In the same way, if the wind is blowing from the right the danger is from the left and from behind, and if blowing from the left the danger is from the right and from behind. In none of these cases is there any appreciable danger of an attack from in front, for in my experience all unwounded tigers, whether man-eaters or not, are disinclined to make a head-on attack. Under normal conditions man-eating tigers limit the range of their attack to the distance they can spring, and for this reason they are more difficult to cope with than wounded tigers, who invariably launch an attack from a little distance, maybe only ten or twenty yards, but possibly as much as a hundred yards. This means that whereas the former have to be dealt with in a matter of split seconds, the latter give one time to raise a rifle and align the sights. In either case it means rapid shooting and a fervent prayer that an ounce or two of lead will stop a few hundred pounds of muscle and bone.

In the case of the tigress I was hunting, I knew that her wound would not admit of her springing and that if I kept out of her reach I would be comparatively safe. The possibility that she had recovered from her wound in the four days that had elapsed since I had last seen her had, however, to be taken into account. When therefore I started out alone on the morning of 11 April to take up the tracks where I had left them the previous evening, I resolved to keep clear of any rock, bush, tree, or other object behind which the tigress might be lying up in wait for me.

She had been moving the previous evening in the direction of the Tanakpur road. I again found where she had spent the night, this time on a soft bed of dry grass, and from this point I followed her fresh tracks. Avoiding dense cover—possibly because she could not move through it silently—she was keeping to water channels and game tracks and it became apparent that she was not moving about aimlessly but was looking for something to kill and eat. Presently, in one of these water channels she found and killed a few-weeks-old *kakar*. She had come on the young deer as it was lying asleep in the sun on a bed of sand, and had eaten every scrap of it, rejecting nothing but the tiny hooves. I was now only a minute or two behind her, and knowing that the morsel would have done no more than whet her appetite, I redoubled my precautions. In places the channels and game tracks to which the tigress was keeping twisted and turned and ran through dense cover or past rocks. Had my condition been normal I would have followed on her footsteps and possibly been able to catch up with her, but unfortunately I was far from normal. The swelling on my head, face, and neck, had now increased to such proportions that I was no longer able to move my head up or down or from side to side, and my left eye was closed. However, I still had one good eye, fortunately my right one, and I could still hear a little.

During the whole of that day I followed the tigress without seeing her and without, I believe, her seeing me. Where she had gone along water channels, game tracks, or cattle paths that ran through dense cover I skirted round the cover and picked up her pug-marks on the far side. Not knowing the ground was a very great handicap, for not only did it necessitate walking more miles than I need have done, but it also prevented my anticipating the movements of the tigress and ambushing her. When I finally gave up the chase for the day, the tigress was moving up the valley in the direction of the village.

Back in camp I realized that the 'bad time' I had foreseen and dreaded was approaching. Electric shocks were stabbing through the enormous abscess, and the hammer blows were increasing in intensity. Sleepless nights and a diet of tea had made a coward of me, and I could not face the prospect of sitting on my bed through another long night, racked with pain and waiting for something, I knew not what, to happen.

I had come to Talla Des to try to rid the hill people of the terror that menaced them and to tide over my bad time, and all that I had accomplished so far was to make their condition worse. Deprived of the ability to secure her natural prey, the tigress, who in eight years had only killed a hundred and fifty people would now, unless she recovered from her wound, look to her easiest prey—human beings—to provide her with most of the food she needed. There was therefore an account to be settled between the tigress and myself, and that night was as suitable a time as any to settle it.

Calling for a cup of tea—made hill-fashion with milk—which served me as dinner, I drank it while standing in the moonlight. Then, calling my eight men together, I instructed them to wait for me in the village until the following evening, and if I did not return by then to pack up my things and start early the next morning for Naini Tal. Having done this I picked up my rifle from where I put it on my bed, and headed down the valley. My men, all of whom had been with me for years, said not a word either to ask me where I was going or to try to dissuade me from going. They just stood silent in a group and watched me walk away. Maybe the glint I saw on their cheeks was only imagination, or maybe it was only the reflection of the moon. Anyway, when I looked back not a man had moved. They were just standing in a group as I had left them.

8

One of my most pleasant recollections—of the days when I was young—are the moonlight walks along forest roads that ten or a dozen of us used to take during the winter months, and the high teas we consumed on our return home. These walks tended to dispel all the fears that assail a human being in a forest at night, and, further, they made us familiar with the sounds to be heard in a forest by night. Later, years of experience added to my confidence and to my knowledge. When therefore I left my camp on the night of 11 April—in brilliant moonlight—to try conclusions with the Talla Des man-eating tigress, I did not set out with any feeling of inferiority on what might appear to have been a suicidal quest.

I have been interested in tigers from as far back as I can remember, and having spent most of my life in an area in which they were plentiful I have had ample opportunities of observing them. My ambition when I was very young was to see a tiger, just that, and no more. Later my ambition was to shoot a tiger, and this I accomplished on foot with an old army rifle which I bought for fifty rupees from a seafaring man, who I am inclined to think had stolen it and converted it into a sporting rifle. Later still, it was my ambition to photograph a tiger. In the course of time all three of these ambitions were fulfilled. It was while trying to photograph tigers that I learnt the little I know about them. Having been favoured by Government with the 'freedom of the forests', a favour which I very greatly appreciate and which I shared with only one other sportsman in India, I was able to move about without let or hindrance in those forests in which tigers were most plentiful. Watching tigers for days or weeks on end, and on one occasion for four and a half months, I was able to learn a little about their habits and in particular their method of approaching and of killing their victims. A tiger does not run down its prey; it either lies in wait or stalks it. In either case contact with its victim is made by a single spring, or by a rush of a few yards followed by a spring. If therefore an animal avoids passing within striking distance of a tiger, avoids being stalked, and reacts instantly to danger whether conveyed by sight, scent, or by hearing, it has a reasonable chance of living to an old age. Civilization has deprived human beings of the keen sense of scent and hearing enjoyed by animals, and when a human being is menaced by a man-eating tiger he has to depend for his safety almost entirely on sight. When restlessness and pain compelled me to be on the move that night, I was handicapped to the extent that I only had one effective eye. But against this handicap was the knowledge that the tigress could do me no harm if I kept out of her reach, whereas I could kill her at a distance. My instructions therefore to my men to go back to Naini Tal if I failed to return by the following evening, were not given because I thought I could not cope with the tigress, but because I feared there was a possibility of my becoming unconscious and unable to defend myself.

One of the advantages of making detailed mental maps of ground covered is that finding the way back to any given spot presents no difficulty. Picking up the pug-marks of my quarry where I had left them, I resumed my tracking, which was now only possible on game tracks and on cattle paths, to which the tigress was, fortunately, keeping. *Sambhar* and *kakar* had now come out on to the open glades, some to feed and others for protection, and though I could not pin-point their alarm calls they let me know when the tigress was on the move and gave me a rough idea of the direction in which she was moving.

On a narrow, winding cattle path running through dense cover I left the pug-marks of the tigress and worked round through scattered brushwood to try to pick them up on the far side. The way round was longer than I had anticipated, and I eventually came out on an open stretch of ground with short grass and dotted about with big oak trees. Here I came to a halt in the shadow of a big tree. Presently, by a movement of this shadow, I realized that the tree above me was tenanted by a troop of *langurs*. I had covered a lot of ground during the eighteen hours I had been on my feet that day, and here now was a safe place for me to rest awhile, for the *langurs* above would give warning of danger. Sitting with my back against the tree and facing the cover round which I had skirted, I had been resting for half an hour when an old *langur* gave his alarm call; the tigress had come out into the open and the *langur* had caught sight of her. Presently I, too, caught sight of the tigress just as she started to lie down.

She was a hundred yards to my right and ten yards from the cover, and she lay down broadside on to me with her head turned looking up at the calling *langur*.

I have had a lot of practice in night shooting, for during the winter months I assisted our tenants at Kaladhungi to protect their crops against marauding animals such as pig and deer. On a clear moonlight night I can usually count on hitting an animal up to a range of about a hundred yards. Like most people who have taught themselves to shoot, I keep both eyes open when shooting. This enables me to keep the target in view with one eye, while aligning the sights of the rifle with the other. At any other time I would have waited for the tigress to stand up and then fired at her, but unfortunately my left eye was now closed and a hundred yards was too far to risk a shot with only one eye. On the two previous nights the tigress had lain in the one spot and had possibly slept most of the night, and she might do the same now. If she lay right down on her side—she was now lying on her stomach with her head up—and went to sleep I could either go back to the cattle path on which I had left her pug-marks and follow her tracks to the edge of the cover and get to within ten yards of her, or I could creep up to her over the open ground until I got close enough to make sure of my shot. Anyway, for the present I could do nothing but sit perfectly still until the tigress made up her mind what she was going to do.

For a long time, possibly half an hour or a little longer, the tigress lay in the one position, occasionally moving her head from side to side, while the old *langur* in a sleepy voice continued to give his alarm call. Finally she got to her feet and very slowly and very painfully started to walk away to my right. Directly in the line in which she was going there was an open ravine ten to fifteen feet deep and twenty to twenty-five yards wide, which I had crossed lower down when coming to the spot where I now was. When the tigress had increased the distance between us to a hundred and fifty yards, and the chances of her seeing me had decreased, I started to follow her. Slipping from tree to tree, and moving a little faster than she, I reduced her lead to fifty yards by the time she reached the edge of the ravine. She was now in range, but was standing in shadow, and her tail end was a very small mark to fire at. For a long and anxious minute she stood in the one position and then, having made up her mind to cross the ravine, very gently went over the edge.

As the tigress disappeared from view I bent down and ran forward on silent feet. Bending my head down and running was a very stupid mistake for me to have made, and I had only run a few yards when I was overcome by vertigo. Near me were two oak saplings, a few feet apart and with interlaced branches. Laying down my rifle I climbed up the saplings to a height of ten or twelve feet. Here I found a branch to sit on, another for my feet, and yet other small branches for me to rest against. Crossing my arms on the branches in front of me, I laid my head on them, and at that moment the abscess burst, not into my brain as I feared it would, but out through my nose and left ear.

'No greater happiness can man know, than the sudden cessation of great pain', was said by someone who had suffered and suffered greatly, and who knew the happiness of sudden relief. It was round about midnight when relief came to me, and the grey light was just beginning to show in the east when I raised my head from my crossed arms. Cramp in my legs resulting from my having sat on a thin branch for four hours had roused me, and for a little while I did not know where I was or what had happened to me. Realization was not long in coming. The great swelling on my head, face, and neck had gone and with it had gone the pain. I could now move my head as I liked, my left eye was open, and I could swallow without discomfort. I had lost an opportunity of shooting the tigress, but what did that matter now, for I was over my bad time and no matter where or how far the tigress went I would follow her, and sooner or later I would surely get another chance.

When I last saw the tigress she was heading in the direction of the village. Swinging down from the saplings, up which I had climbed with such difficulty, I retrieved my rifle and headed in the same direction. At the stream I stopped and washed and cleaned myself and my clothes as best I could. My men had not spent the night in the village as I had instructed them to, but had sat round a fire near my tent keeping a kettle of water on the boil. As, dripping with water, they saw me coming towards them they sprang up with a glad cry of 'Sahib! Sahib! You have come back, and you are well.' 'Yes,' I answered, 'I have come back, and I am now well.' When an Indian gives his loyalty, he gives it unstintingly and without counting the cost. When we arrived at Talla Kote the headman put two rooms at the disposal of my men, for it was dangerous to sleep anywhere except behind locked doors. On this my bad night, and fully alive to the danger, my men had sat out in the open in case they could be of any help to me, and to keep a kettle on the boil for my tea—if I should return. I cannot remember if I drank the tea, but I can remember my shoes being drawn off by willing hands, and a rug spread over me as I lay down on my bed.

Hours and hours of peaceful sleep, and then a dream. Someone was urgently calling me, and someone was as urgently saying I must not be disturbed. Over and over again the dream was repeated with slight variations, but with no less urgency, until the words penetrated through the fog of sleep and became a reality. 'You *must* wake him or he will be very angry.' And the rejoinder, 'We will *not* wake him for he is very tired.' Ganga Ram was the last speaker, so I called out and told him to bring the man to me. In a minute my tent was besieged by an excited throng of men and boys all eager to tell me that the man-eater had just killed six goats on the far side of the village. While pulling on my shoes I looked over the throng and on seeing Dungar Singh, the lad who was with me when I shot the cubs, I asked him if he knew where the goats had been killed and if he could take me to the spot. 'Yes, yes,' he answered eagerly, 'I know where they were killed and I can take you there.' Telling the head-man to keep the crowd back, I armed myself with my .275 rifle and, accompanied by Dungar Singh, set off through the village.

My sleep had refreshed me, and as there was now no need for me to put my feet down gently—to avoid jarring my head—I was able for the first time in weeks to walk freely and without discomfort.

9

The day I arrived at Talla Kote, Dungar Singh, the lad who was with me now, had taken me through the village to a narrow saddle from where there was an extensive view into two valleys. The valley to the right fell steeply away in the direction of the Kali river. At the upper end of this valley I had shot the cubs and wounded the tigress. The other valley, the one to the left, was less steep and from the saddle a goat track ran down into it. It was in this valley that the goats had been killed. Down the goat track the lad now started to run, with me close on his heels. After winding down over steep and broken ground for five or six hundred yards, the track crossed a stream and then continued down the valley on the left bank. Close to where the track crossed the stream there was an open bit of comparatively flat ground. Running from left to right across this open ground was a low ridge of rock, on the far side of which was a little hollow, and lying in the hollow were three goats.

On the way down the hill the lad had told me that round about midday a large flock of goats in charge of ten or fifteen boys was feeding in the hollow, when a tiger—which they suspected was the man-eater—suddenly appeared among them and struck down six goats. On seeing the tiger the boys started yelling and were joined by some men collecting

firewood near by. In the general confusion of goats dashing
about and human beings yelling, the tiger moved off and no
one appeared to have seen in which direction it went. Grab-
bing hold of three dead goats the men and boys dashed back
to the village to give me the news, leaving three goats with
broken backs in the hollow.

That the killer of the goats was the wounded man-eater
there could be no question, for when I last saw her the
previous night she was going straight towards the village.
Further, my men told me that an hour or so before my
return to camp a *kakar* had barked near the stream, a
hundred yards from where they were sitting, and thinking
that the animal had barked on seeing me they had built up
the fire. It was fortunate that they had done so, for I later
found the pug-marks of the tigress where she had skirted
round the fire and had then gone through the village,
obviously with the object of securing a human victim. Having
failed in her quest she had evidently taken cover near the
village, and at the first opportunity of securing food had
struck down the goats. This she had done in a matter of
seconds, while suffering from a wound that had made her
limp badly.

As I was not familiar with the ground, I asked Dungar
Singh in which direction he thought the tigress had gone.
Pointing down the valley he said she had probably gone in
that direction, for there was heavy jungle farther down.
While I was questioning him about this jungle, with the idea
of going down and looking for the tigress, a *kalege* pheasant
started chattering. On hearing this the lad turned round and
looked up the hill, giving me an indication of the direction
in which the bird was calling. To our left the hill went up
steeply, and growing on it were a few bushes and stunted
trees. I knew the tigress would not have attempted to climb
this hill, and on seeing me looking at it Dungar Singh said the
pheasant was not calling on the hill but in a ravine round the
shoulder of it. As we were not within sight of the pheasant,

M

there was only one thing that could have alarmed it, and that was the tigress. Telling Dungar Singh to leave me and run back to the village as fast as he could go, I covered his retreat with my rifle until I considered he was clear of the danger zone and then turned round to look for a suitable place in which to sit.

The only trees in this part of the valley were enormous pines which, as they had no branches for thirty or forty feet, it would be quite impossible to climb. So of necessity I would have to sit on the ground. This would be all right during daylight, but if the tigress delayed her return until nightfall, and preferred human flesh to mutton, I would need a lot of luck to carry me through the hour or two of darkness before the moon rose.

On the low ridge running from left to right on the near side of the hollow was a big flat rock. Near it was another and smaller one. By sitting on this smaller rock I found I could shelter behind the bigger, exposing only my head to the side from which I expected the tigress to come. So here I decided to sit. In front of me was a hollow some forty yards in width with a twenty-foot-high bank on the far side. Above this bank was a ten- to twenty-yard-wide flat stretch of ground sloping down to the right. Beyond this the hill went up steeply. The three goats in the hollow, which were alive when the boys and men ran away, were now dead. When striking them down the tigress had ripped the skin on the back of one of them.

The *kalege* pheasant had now stopped chattering, and I speculated as to whether it had called at the tigress as she was going up the ravine after the lad and I had arrived or whether it had called on seeing the tigress coming back. In the one case it would mean a long wait for me, and in the other a short one. I had taken up my position at 2 p.m., and half an hour later a pair of blue Himalayan magpies came up the valley. These beautiful birds, which do a lot of destruction in the nesting season among tits and other small birds, have an uncanny instinct for finding in a jungle anything that is dead. I heard the magpies long before I saw them, for they are very vocal. On catching sight of the goats they stopped chattering and very cautiously approached. After several false alarms they alighted on the goat with the ripped back and started to feed. For some time a king vulture had been quartering the sky, and now, on seeing the magpies on the goat, he came sailing down and landed as lightly as a feather on the dead branch of a pine tree. These king vultures with their white shirt-fronts, black coats, and red heads and legs, are always the first of the vultures to find a kill. Being smaller than other vultures it is essential for them to be first at the table, for when the others arrive they have to take a back seat.

I welcomed the vulture's coming, for he would provide me with information I lacked. From his perch high up on the pine tree he had an extensive view, and if he came down and joined the magpies it would mean that the tigress had gone, whereas if he remained where he was it would mean that she was lying up somewhere close by. For the next half hour the scene remained unchanged—the magpies continued to feed, and the vulture sat on the dead branch—and then the sun was blotted out by heavy rain-clouds. Shortly after, the *kalege* pheasant started chattering again and the magpies flew screaming down the valley. The tigress was coming, and here, sooner than I had expected, was the chance of shooting her that I had lost the previous night when overcome by vertigo.

A few light bushes on the shoulder of the hill partly obstructed my view in the direction of the ravine, and presently through these bushes I saw the tigress. She was coming, very slowly, along the flat bit of ground above the twenty-foot-high bank and was looking straight towards me. With only head exposed and my soft hat pulled down to my eyes, I knew she would not notice me if I made no movement. So, with the rifle resting on the flat rock, I sat perfectly still. When she had come opposite to me the tigress sat down, with the bole of a big pine tree directly between us. I could see her head on one side of the tree and her tail and part of her hindquarters on the other. Here she sat for minutes, snapping at the flies that, attracted by her wound, were tormenting her.

10

Eight years previously, when the tigress was a comparatively young animal, she had been seriously injured in an encounter with a porcupine. At the time she received this injury she may have had cubs, and unable for the time being to secure her natural prey to feed herself in order to nourish her cubs, she had taken to killing human beings. In doing this she had committed no crime against the laws of Nature. She was a carnivorous animal, and flesh, whether human or animal, was the only food she could assimilate. Under stress of circumstances an animal, and a human being also, will eat food that under normal conditions they are averse to eating. From the fact that during the whole of her man-eating career the tigress had only killed a hundred and fifty human beings—fewer than twenty a year—I am inclined to think that she only resorted to this easily procured form of food when she had cubs and when, owing to her injury, she was unable to get the requisite amount of natural food needed to support herself and her family.

The people of Talla Des had suffered and suffered grievously from the tigress, and for the suffering she had inflicted she was now paying in full. To put her out of her misery I several times aligned the sights of my rifle on her head, but the light, owing to the heavy clouds, was not good enough for me to make sure of hitting a comparatively small object at sixty yards.

Eventually the tigress stood up, took three steps and then stood broadside on to me, looking down at the goats. With my elbows resting on the flat rock I took careful aim at the

spot where I thought her heart would be, pressed the trigger,
and saw a spurt of dust go up on the hill on the far side of her.
On seeing the dust the thought flashed through my mind
that not only had I missed the tigress's heart, but that I had
missed the whole animal. And yet, after my careful aim, that
could not be. What undoubtedly had happened was that my
bullet had gone clean through her without meeting any
resistance. At my shot the tigress sprang forward, raced over
the flat ground like a very frightened but unwounded animal,
and before I could get in another shot disappeared from view.

Mad with myself for not having killed the tigress when she
had given me such a good shot, I was determined now that
she would not escape from me. Jumping down from the
rock, I sprinted across the hollow, up the twenty-foot bank
and along the flat ground until I came to the spot where the
tigress had disappeared. Here I found there was a steep
forty-foot drop down a loose shale scree. Down this the
tigress had gone in great bounds. Afraid to do the same for
fear of spraining my ankles, I sat down on my heels and
tobogganed to the bottom. At the foot of the scree was a
well-used footpath, along which I felt sure the tigress had
gone, though the surface was too hard to show pug-marks.
To the right of the path was a boulder-strewn stream, the
one that Dungar Singh and I had crossed farther up, and
flanking the stream was a steep grassy hill. To the left of the
path was a hill with a few pine trees growing on it. The path
for some distance was straight, and I had run along it for
fifty or more yards when I heard a *ghooral* give its alarm
sneeze. There was only one place where the *ghooral* could be
and that was on the grassy hill to my right. Thinking that the
tigress had possibly crossed the stream and gone up this hill,
I pulled up to see if I could see her. As I did so, I thought I
heard men shouting. Turning round I looked up in the
direction of the village and saw a crowd of men standing on
the saddle of the hill. On seeing me look round they shouted
again and waved me on, *on*, straight along the path. In a

moment I was on the run again, and on turning a corner found fresh blood on the path.

The skin of animals is loose. When an animal that is standing still is hit in the body by a bullet and it dashes away at full speed, the hole made in the skin does not coincide with the hole in the flesh, with the result that, as long as the animal is running at speed, little if any blood flows from the wound. When, however, the animal slows down and the two holes come closer together, blood flows and continues to flow more freely the slower the animal goes. When there is any uncertainty as to whether an animal that has been fired at has been hit or not, the point can be very easily cleared up by going to the exact spot where the animal was when fired at, and looking for cut hairs. These will indicate that the animal was hit, whereas the absence of such hairs will show that it was clean missed.

After going round the corner the tigress had slowed down, but she was still running, as I could see from the blood splashes, and in order to catch up with her I put on a spurt. I had not gone very far when I came to a spur jutting out from the hill on my left. Here the path bent back at a very acute angle, and not being able to stop myself, and there being nothing for me to seize hold of on the hillside, I went over the edge of the narrow path, all standing. Ten to fifteen feet below was a small rhododendron sapling, and below the sapling a sheer drop into a dark and evil-looking ravine where the stream, turning at right angles, had cut away the toe of the hill. As I passed the sapling with my heels cutting furrows in the soft earth, I gripped it under my right arm. The sapling, fortunately, was not uprooted, and though it bent it did not break. Easing myself round very gently, I started to kick footholds in the soft loamy hill-face which had a luxuriant growth of maidenhair fern.

The opportunity of catching up with the tigress had gone, but I now had a well-defined blood-trail to follow, so there was no longer any need for me to hurry. The footpath which at first had run north now ran west along the north face of a steep and well-wooded hill. When I had gone for another two hundred yards along the path, I came to flat ground on a shoulder of the hill. This was the limit I would have expected a tiger shot through the body to have travelled, so I approached the flat ground, on which there was a heavy growth of bracken and scattered bushes, very cautiously.

A tiger that has made up its mind to avenge an injury is the most terrifying animal to be met with in an Indian jungle. The tigress had a very recent injury to avenge and she had demonstrated—by striking down six goats and by springing and dashing away when I fired at her—that the leg wound she had received five days before was no handicap to rapid movement. I felt sure, therefore, that as soon as she became aware that I was following her and she considered that I was within her reach, she would launch an all-out attack on me, which I would possibly have to meet with a single bullet. Drawing back the bolt of the rifle, I examined the cartridge very carefully, and satisfied that it was one of a fresh lot I had recently got from Manton in Calcutta, I replaced it in the chamber, put back the bolt, and threw off the safety catch.

The path ran through the bracken, which was waist high and which met over it. The blood trail led along the path into the bracken, and the tigress might be lying up on the path or on the right or the left-hand side of it. So I approached the bracken foot by foot and looking straight ahead for, on these occasions, it is unwise to keep turning the head. When I was within three yards of the bracken I saw a movement a yard from the path on the right. It was the tigress gathering herself together for a spring. Wounded and starving though she was, she was game to fight it out. Her spring, however, was never launched, for, as she rose, my first bullet raked her from end to end, and the second bullet broke her neck.

Days of pain and strain on an empty stomach left me now trembling in every limb, and I had great difficulty in reaching the spot where the path bent back at an acute angle and where, but for the chance dropping of a rhododendron seed, I would have ended my life on the rocks below.

The entire population of the village, plus my own men, were gathered on the saddle of the hill and on either side of it, and I had hardly raised my hat to wave when, shouting at the tops of their voices, the men and boys came swarming down. My six Garhwalis were the first to arrive. Congratulations over, the tigress was lashed to a pole and six of the proudest Garhwalis in Kumaon carried the Talla Des man-eater in triumph to Talla Kote village. Here the tigress was laid down on a bed of straw for the women and children to see, while I went back to my tent for my first solid meal in many weeks. An hour later with a crowd of people round me I skinned the tigress.

My first bullet, a .275 soft-nose with split nickel case fired on 7 April, was bushed and firmly fixed in the ball-and-socket joint of the tigress's right shoulder. The second and third bullets, fired as she was falling through the air and climbing up the hill, had missed her. The fourth, fired on 12 April, had gone clean through without striking any bones, and the fifth and sixth had killed her. From her right foreleg and shoulder I took some twenty porcupine quills, ranging in length from two to six inches, which were firmly embedded in muscle and were undoubtedly the cause of the tigress's having become a man-eater.

I spent the following day in partly drying the skin, and three days later I was safely back in my home with my bad time behind me. Baynes very kindly sent for Dungar Singh and his brother, and at a public function at Almora thanked

them for the help they had given me and presented them
with my token of gratitude. A week after my return to Naini
Tal, Sir Malcolm Hailey gave me an introduction to Colonel
Dick, an ear specialist, who treated me for three months in
his hospital in Lahore and restored my hearing sufficiently
for me to associate with my fellow men without embarrassment, and gave me back the joy of hearing music and the
song of birds.

Epilogue

THE story of the Talla Des man-eater—which I refrained from telling until I had written *Jungle Lore*—has now been told. I am aware that to many the story will seem incredible, and to none more so than to those who have themselves hunted tigers. None knows better than I that the hunting of tigers on foot is not a popular sport, and that the hunting of man-eaters on foot is even less so. I also know that the following-up of a wounded tiger on foot is a task that is sought by none and dreaded by all. And yet, knowing these things, I have told of the hunting of a man-eating tiger on foot, not only by day but also by night, and the chasing on foot of a wounded tiger. Small wonder, then, if my story to many should seem incredible.

There are few places in Kumaon where a fortnight's holiday could be more pleasantly spent than along the eastern border of the Almora district. Hiking in the Himalayas is becoming a very popular pastime, and I could suggest no more pleasant hike for a sportsman or for a party of young army men or students than the following:

Start from Tanakpur, but before doing so get the *peshkar* to give you a *tahsil peon* to show you where the epic fight took place between the elephant and the two tigers. From Tanakpur go via Baramdeo to Purnagiri. Here, after doing *darshan* at the temple, learn all you can from the High Priest and the temple *pujaris* about the lights that appear on the far side of the Sarda, and similar manifestations, as for example the fire with an old man sitting near it telling his beads that is to be seen at certain seasons at the foot of the Pindari glacier. From Purnagiri a track used by the priests will take you to Thak village. This is beautifully situated, and while you rest and admire the view, get the headman or any of the other men sitting round to give you his version of the shooting of the Thak and of the Chuka man-eating tigers. Tewari, a relative of the headman and as fine a type of a hill Brahmin as you will see, will then show you where his brother whose body he helped me to find was killed, the mango tree with a spring at its roots, and the rock on the way down to Chuka where I shot the Thak man-eater. He will also, if you have the time, show you the *ficus* tree from which I shot the Chuka man-eater. At Chuka inquire for Kunwar Singh, and hear his story of the hunting of the two tigers.

From Chuka to Talla Kote is a long march, and it will be advisable to start at crack of dawn. Having forded the Ladhya near its junction with the Sarda, you will come to Sem. The headman of Sem, who was a boy when I knew him, will show you where the man-eater killed his mother while she was cutting grass near their home. With Sem behind you and a stiff climb accomplished, you will pass the small hamlet where I spent a night under a mango tree. After going over the ridge you will come to a forest road. Take the turn to the left and follow the road until you come to a stream. Cross the stream and the small patch of open ground on which my 40-lb. tent was pitched, and you have reached your destination, Talla Kote.

Dungar Singh, *malguzar* (land-holder) of Talla Kote, will now be about forty years of age. Give him my *salams* and ask him to take you to the *ling* or saddle from which there is an extensive view into two valleys. Face first the valley to the east and get Dungar Singh to point out the bush where his mother was killed, the oak tree under which she was eaten, the *wyran* field on which the young tigers were shot, and the grassy hill up which the wounded tigress went. Then turn round, walk a few steps, and face the valley to the west. Dungar Singh will now point out where the six goats were killed, where the tigress was standing when my bullet went through her, and the footpath along which she dashed and along which I ran after her.

The hunting of no other tiger has ever been witnessed by a greater number of non-participants than witnessed the hunting of the Talla Des man-eating tiger. Some of those will have passed away, but many will still remain and they will not have forgotten my visit or the thrilling events of the week I spent with them.

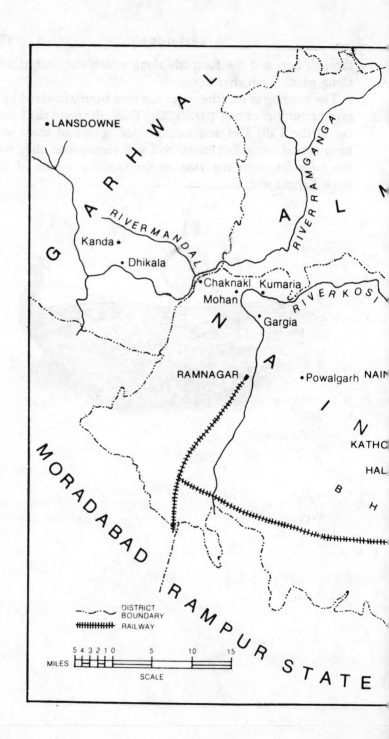

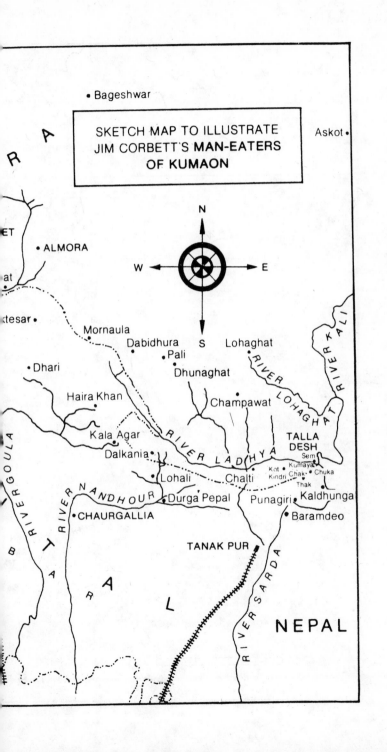